The Technology Used By The Christ

Lawrence Hazlewood

Order this book online at www.trafford.com/07-0447
or email orders@trafford.com

Most Trafford titles are also available at major online book retailers.

Note for Librarians: A cataloguing record for this book is available from Library and Archives Canada at www.collectionscanada.ca/amicus/index-e.html

ISBN: 978-1-4251-2043-6

We at Trafford believe that it is the responsibility of us all, as both individuals and corporations, to make choices that are environmentally and socially sound. You, in turn, are supporting this responsible conduct each time you purchase a Trafford book, or make use of our publishing services. To find out how you are helping, please visit www.trafford.com/responsiblepublishing.html

Our mission is to efficiently provide the world's finest, most comprehensive book publishing service, enabling every author to experience success. To find out how to publish your book, your way, and have it available worldwide, visit us online at www.trafford.com/10510

www.trafford.com

North America & international
toll-free: 1 888 232 4444 (USA & Canada)
phone: 250 383 6864 ♦ fax: 250 383 6804
email: info@trafford.com

The United Kingdom & Europe
phone: +44 (0)1865 722 113 ♦ local rate: 0845 230 9601
facsimile: +44 (0)1865 722 868 ♦ email: info.uk@trafford.com

10 9 8 7 6 5 4 3 2

THE TECHNOLOGY USED BY THE CHRIST

The greatest cultural shock since the world became round

This technology has been beyond the understanding of mankind for many thousands of years although it surprisingly has been kept and passed on through folklore and dogma. Until the advent and general understanding of atomic and advanced electronic technology became available it has not been of any use to the population at large. Therefore the content of this essay must cover the actuality, (things which are in full existence) atom, electronics, speech, belief systems, thought, vibration, all of humanity, real, reality, fact, God, religion, spirit, mental manipulation of atomic Structure, commerce, finance, history, life, the universe and the whole of the things within the universe.

It is hoped this will give the reader a clearer understanding of the pre designed purpose for the human race and the difference between commerce and the world of dogma we are expected to live in, and the part of you which are the least known but are parts that are referred to as the ones which are more than the total sum of your body's parts.

The development of these parts are the only opportunity for you to reach the potential the human is born to achieve, for it will give to you all truth, all power, all knowledge, all wealth, all abundance, total freedom, all creative power.

INTRODUCTION

A thought went through my mind which said, in the hands of the stubborn egotistic greedy minds of man, religion in the name of God has created the most bloody three or four thousand years this planet has ever known. To commerce I am a statistic which creates financial wealth for the wealthy. To my nation I am in the first half of my life gun fodder. In the second half of life I am being prepared for the final insult of old age whereby I can then be legally discriminated against by the people who have required my financial help to retain their life style.

The Bible tells me I am greater than my total number of my parts, and that I can have everlasting life, that I can rise above the trials and tribulations which are inflicted by the few upon the many, also that I am journeying toward home.

To my elected representative I am something to be manipulated, so, who am I? What am I? From whence did I come? Where am I going? Is there a way? Is there something better? What is life for? Am I doing it right? Am I hearing and reading the truth? What is all this pain and suffering about? Who brings it on and why? What are miracles? How are they performed? Where the hell am I and what is my place in the cosmos?

In that moment it became clear to me the difference between a thinker of the species man, and a dreamer drifting along on what they imagine to be correct. It was only a matter of values and how you interpret them, what you depend on for you own personal interpretation, and how desperate is the need of others, for you to follow their interpretation of understanding.

In making a conscious decision to delve into the original thoughts and pursue them to the end, I decided to start as early in history as possible.

My findings show the Bible gives us the creation, and the never ending process of that creation, but the truth of it has been absolutely shamefully distorted This of course started other questions.

Why? By whom? For what purpose? Is there a system of God, and a system of man? What is commerce? What is religion? What is the church position? Are the demonstrations shown us by Jesus being put into action? Is the truth of any such action being withheld? What are the lies? Why are there liars? What is the truth? Do we believe the truth? How would we handle the truth?

Without going into every nook and cranny of this world's history, I will explain the important parts, and leave the argument and deceit for those who would keep the status quo for whatever reason

After much deliberation I came to the conclusion that if you really believe a lie, that lie becomes your fact. Circumstantial information. Then from that fact you make all your judgments and assumptions, irrespective of how terribly wrong those judgments and assumptions may be, and the trouble they may bring to yourself or others. A wrong decision carries the seed of disaster.

Writings tell of a civilization which inhabited planet earth some time before Lemuria. The people of this civilization lived according to the ideal principle of the universe therefore they lived in what the present civilization would imagine to be heaven.

Following these thoughts it appeared to me that the times have not changed much in many hundreds of years. It seems the collective intelligence of humanity has not kept pace with the commercial advances, and that the spin doctors who make the most noise are the ones who have been responsible for the almost non existence of humanities spiritual development.

Just take a peep into the past of the great discoveries of history and the people who have bought them to the notice of the world at large. Also consider the penalty that was extracted from them for speaking out above the supposed knowledge of the established order. Right up until the sixteen hundreds the manner used to keep the status quo and silence those who had the ability to think out side of the box, was to ridicule that person and under some pretense have them put to death.

Strangely enough the information put before the people in this manner never did die with the one who had the original thought. Just as well for the world would still be flat. The Spanish inquisition would still be with us. The sun would still be going around the world. The law of water displacement would not be available to mathematics. plus the hundreds of other things discovered by people who were never in the selected group, known as the wise men of the time.

Little did the supposed wise men of that or present time know that it was not the intention of the thinker to expose the limit of the wise man's knowledge, that it was only to improve humanities general understanding. The only exposure of the wise mens lack of knowledge was there own exposure of their own ego, endeavoring to suppress new information which ended in the death of the thinker so they could maintain the status quo.

I think this still goes on in this day and age as the established groups are very quick to publicly ridicule any who do not have the piece of paper to say they have learned all that is known of any subject at the time the piece of paper was written. They seem very reluctant to give credit to any who endeavor to bring new thought into any subject although they would never think of themselves as bigots.

It is unfortunate that the ego of any established group can not understand the in depth communication of the new theory and devote time in serious investigation to bring about a satisfactory win win situation for all humanity .

The beginning of this system was created when man stopped being nomadic and started to till the soil. The reasoning used is still the same and is that of fear of the loss of the material and physical. It has built on itself for ten thousand years and is kept in place by those who think they have the most to lose financially or the loss of power over the people they are only there to represent.

As time has past the reasoning becomes stronger and the outcomes returned from the planting of a seed of deceit, dishonesty, and corruption have for that last ten thousand years brought about wholesale slaughter through war down to the neighbours disputes.

This system is even at this time using entertainment to plant the seeds of sex, murder, violence, hate, fear, in the youngest of the world's population and is making sure the fertility of the nurturing arena of the seed is supplied with all that will help the seed to mature. It makes me wonder if this action could be the measure of the developing intelligence of the specie man, or a well established beginning of the dissolution of the period of control we have been under these last ten thousand years. Is man waking up? Has he really had enough? Has he had enough to peacefully force changes?

All this time and particularly in the last two thousand years, the offer of a wondrous life through the inner discovery of spirit in self has been left largely unexplored until the present day.

The true intelligence of the people will decide whether or not the time has come to re-evaluate the belief systems we have mostly been compelled to use as our basis of fact, or whether the capability of the mind to expand its conscience is the way of the future.

Well, there needs to be a beginning so here we go.

Lawrence Hazlewood.

The history of humanity on this planet including the intergalactic biological genetic experimentation has kept on repeating itself for I shudder to thing how many centuries.

There were certainly civilizations before Lemuria, but the knowledge of these is very remote, so the time of their existence is very hazy. Lemuria, and the fact that it did exist, grow to a high standard of technology then disappeared, is no more than what happened to Atlantis and the ones that preceded MU.

From these there is always some people and artifacts which survive to inform those who follow of their developments and downfall. When folklore or any identification is found which does not suit the established academics of the time, it is not recognized, or eluded to as an artifact of some less advanced religious right.

In many instances its significance is totally misunderstood by the scholars, as they seem to mistakenly imagine our present technology has reached the highest standard ever obtained in this universe. Therefore all judgments are made which require the official stamp of the academics, right or wrong, or the out spoken one feels the sting of those who may be afraid of the truth or their imagined standing in the community.

The very first people on planet earth upheld the tenants of the Divine Principle and therefore lived a bountiful life where they had all they could wish for. That which you and I would call heaven. They could all do the things Jesus demonstrated in his time on earth, and they did not know death.

The pool of knowledge is every thought or word used on this planet, and which is stored within the ionized belts of this planet. That includes what you, I, or anyone has said or thought throughout the entire history of man kind on this planet. It is the records.

It was this civilization which placed the original thoughts in the pool of knowledge, and as they knew no negative thought, they could not speak a negative word. Consequently there could never be a negative situation for they only knew and applied love.

After the destruction of MU, the few remaining residents who had tried to make it a better place for all found themselves on the sea shore of what is now known as Mongolia. They were able to put into practice that which they had tried to help the rest of the population to understand. This gave them a good start, and beneath the sands of the Gobi desert one day will be found cities with evidence a plenty of their golden age.

Again the lesser ideal attitude of the past crept in and their community was devastated once more, but still there remained enough people of that race to start again.

It was the lot of these few and those reincarnated back into their own race, to suffer the misdeeds of all the many years of not learning, and putting into action the spiritual information available to them.

As time went by, the land was lifted by earthquakes until it assumed its present size and contour. That is China and all of Asia including Japan and all offshore islands. Part of M U will one day be found of the coast of Japan.

The few people who started the Asian race were those who were the remnants of MU, and they still do not totally understanding the spiritual calling required of them. They prefer to bow to their human masters demand for total obedience.

Another group of souls materialized in flesh in Israel. These people were known as the light people as they were created by the pure white light vibration. Their mission was to show those left on earth the way back to the understanding they used many years before their fall from grace

This is the people known as the Adamic tribe, some thirty five thousand of them, although through changes in the writings you would know them as .Adam and Eve.

Even in this tribe some preferred the call of the flesh in the place of the ideal principle of the universe, which brought about a drifting away of those who followed the tenants

The split away group were known as the Aryan tribe. They were the ones who retained the true tenets, and to make sure the truth would never be lost they made through the process of thought, twelve sets of tablets, which came into being with only the truth of the most high on them.

These tablets were of white marble with green inlay and are now awaiting discovery in twelve different places throughout the world. They will not have to be deciphered as to those who are sensitive enough they will hear the tablets sing their message.

The crystal skulls were formed in the same manner and contain a similar message, also available to any who is sensitive enough.

This tribe also divided and some moved on. This was preordained of course for their purpose was to spread the truth. In time as this process kept repeating itself the whole of Europe was settled.

In the early part of this process a generally knowledgeable group distanced themselves by a very long distance and founded what is known today as Atlantis. Being able to extend their thought into greater achievements they become what they imagined to be a mighty empire.

Their state of development was only slightly better than we have attained in this year of 2003. They were heavily involved in genetic engineering and had produced the animals with human heads as beasts of burden which could talk. These were stolen life forms.

In the final stages as the continent was splitting up into islands, they were very concerned with their own arrogance which encouraged them to be self centred and war like. .

Unfortunately, through imagining that they were the power they lost sight of their spiritual abilities.

While the power was through them they continued to improve that empire, and as the less understanding self-serving leaders pushed the idea that they were the great ones, the ideal which had made the whole of Atlantis what it was, started to diminish.

From the habitation of Atlantis through its breaking into islands to the final disappearance was a period of around two hundred thousand years.

This present civilization is deemed to have started in the year we call ten thousand B. C. This is still to be proven.

Throughout this time groups of people were leaving Atlantis and this is how the Americas, Pacific islands Australia, and

New Zealand were populated. Easter Island was started with a healthy knowledge of original Atlantis which was also lost, as it was in all these countries, but was retained remarkably well in each countries tribal lore.

Some of the people of Atlantis moved into England where they lost touch and became dormant after Stonehenge. Those who went to Europe did much the same thing, while those who went to North Africa became what were known as the ancient Egyptians.

Osiris was one of these, and it was he who tried, and had some success with the undoing of the Atlantean genetic engineering of animal and human tissue, which resulted in what was known as the things. These were the horses with human heads, mermaids and the like which could all talk and were treated like beasts of burden. This was the reason for the down fall of Atlantis for they crated life, abused it and failed to be responsible for it.

When you create something you are responsible for it, or you and your supporters will be consumed by the immorality of your deeds. That does not mean the elastic morality of the human's interpretation of the word.

Another civilization which was short lived and achieved nothing was destroyed in the great flood at the time of Noah. This was the cleansing which destroyed the genetically engineered things from earth together with the unbelievers of the time. It released the souls of those trapped in the falsity of flesh, animal and human.

This action of the 'all omnipotent wisdom' cleared the way for a new start with the wisdom being explained through Noah

After this the tribes of Europe and the middle east were still largely nomadic, and it was not until about ten thousand years B.C. when man started to cultivate the soil that we have gained a greater understanding of the progress of that which is eluded to as modern man.

The academics of each civilization have run away with the idea they were modern man, but the number of people throughout all time who realized they were unlimited continuous life "spiritual creations" and acted accordingly were very few. That time has not had much meaning for those who have not come to realize their spiritual selves, and followed the old teachings from their earlier incarnations, so, as we are today we are faced with another global disaster. We have not learned to developed our spiritual capabilities so why do we need life.

Egypt's recorded history shows polar changes and the reversal of the east and west no less than four times. About eight hundred to seven hundred B C there were smaller cleansing near twelve years apart, although they still destroyed many thousands of lives. See Isaiah

In forty AD Jesus accepted the responsibility for this planet and all that exists on it, to bring humanity back into the light of its true destiny. He has still allowed freedom of choice trusting each and every person will discover the truth of giving to all, only love,

The difference between you individually as the power, is that the real power is allowed by you, to pass through you, and in the passing be accelerated by you. Then will all become clear. You think not?, then lets take a meaningful look at the history of the last ten thousand years, then into the future.

When the species man started ten thousand years ago to cultivate the soil, he unknowingly opened the door for the first commercialisation of this period of existence, or Epoch or (Better known in universal language where time and space do not exist) as a generation.

As those who were still nomadic would pass by a cultivated area, they would just help themselves to what ever they needed to get them to the next place of abundance. This was fine until the newly settled farmer found there was nothing left for him and his family. The obvious thing followed when he stopped the nomads from helping themselves. The nomads need was enough supplies to get to the next place of plenty, so he had to trade or work for those supplies.

Thus began this epoch's trade with the disaster of ownership, and the beginning of unfair trade which was, and still is greed and domination. This was followed by the accumulation of all sorts of food and goods, and the necessary buildings to house it and guards to watch over it all.

Next came alliances among farmers and the establishment of borders which became countries. Then the army to protect it, and the overseer or Wiseman to control it on behalf of the farmers.

Fear of a take over was always on the agenda, and was kept there by the overseer for the purpose of securing his own position and steadily increasing power.

All this was a consolidation of the overseer's power, talking those he represented into the fear of losing material things with the time worn and still successfully used "it will be better, it will be more efficient, it will be safer, every one will be better off, we will all be able to live in peace etc."

As time went by contact with, and application of spiritual knowledge got lost in the connivance of power and trade.

The few followers of truth were and still are a constant annoyance to the stealers of the peoples power. so to look even greater than they where, the so called leaders called themselves priests, high priests, or wise men. There by endeavoring to equal or better those who had more power and set themselves up as kings.

To complement and satisfy their enormous ego, they used the peoples means of trade and the captive slaves to build the colossal structures they called temples into which they placed their stone creations and called them the Gods which must be appeased. It was deemed that only the high priests were able to interpret the wishes of these stone or metal creations. Same as the church leaders today. (HA Bloody HA! Their own ego blocked any contact with the most high)

To keep the status quo it became convenient to eliminate any opposition by making those who did not agree with temple dictates a sacrifice to the Gods.

During the period 10,000 B.C. to 6,000 B.C. things had fairly consolidated themselves and we find Nippur the Babylonian, the centre of eastern culture.

The day of the academic thinking "it stands to reason, it will be better, you can't possibly do without it, imagined knowledge with out concrete fact of all things known and unknown" had truly arrived.

4,400 B.C. was the time Menes became the first Pharaoh of Egypt.

2,300 B.C. saw the time of Khammurabi, the great Babylonian ruler and giver of material law.

This law and enforcement was unacceptable to humanity, as people of the time no longer wanted to be slaves. This bought about the appearance of the great Hebrew patriarch Abraham in 2,100 B.C. but most of those who needed his ministry rejected it. The language of the day could not explain such high technology and the people did not have the faith to just allow it to happen. You do not need to know all this, just be, implement the highest ideal you can call to mind and your faith will do the rest.

The next person allowing the power to flow through himself and to demonstrate the method to be used was Moses in 1400 B C. Moses also new that the number of people who could follow his demonstrations were few, and that the essence of those demonstrations would in time be lost. He therefore wrote the first five books of the bible for those who wished to follow, to gain the understanding and put it to use. You will find out who and why those five books were changed later.

1300 BC saw the exodus of the Israelites from Egypt and another great cleansing of the earth. As soon as the Israelites had reached safety and the waters closed in behind them the darkness overtook the whole world, and lasted twenty five years. This is the time they spent wandering in the desert, also it was when the honey flavoured manna fell from heaven which was the only sustenance for the few people left world wide After the first fifteen years trees started to show signs of coming into leaf again, then ten years later the sun became visible. 1300 BC was also the death of Moses.

1100 BC and David second king of Israel was born. The story of his rise and fall together with the reasons are well written in the bible, so there is no need for me to comment further, other than to say that, during his time of using God through the universal principle ideal, he did accomplish many great things.

Solomon son of David, who all of his life implemented the principle ideal of the heavens, became great for his unfailing justice in which all people who came before him found themselves in a win-win situation Therefore his cup naturally had to be full and running over, even with material things

Solomon's time on earth finished 937 BC.

753 BC Rome was founded then three years later the Romans seized Sabine woman for wives. Who were the Romans? From whence did they come? Why did they need to seize other woman for wives.?

560 BC saw the birth of Buddha. His philosophy was certainly the way to the light. Alas the knowledge he bought became lost through academic misinterpretation, as has all truth of things spiritual.

The greatest event in the history of this world occurred in 4 BC when the baby Jesus was born, and the greatest disaster happened in 30 AD when he was crucified. Why the high technology Jesus demonstrated was put down along with him and his devoted followers, will be explained in the following pages.

About 150 BC a ruthless and much feared nomadic tribe decided to stop their wandering and settle down They did not have land and were only accepted because of the fear they generated. Part of the idea of settling was to be seen as respectable, although the uppermost idea was to gain control over as much and as many as possible.

When they sat down and discussed the method they would use; it became evident that to offer a great reward and give nothing in return and make a charge for advice was a system that could be used on the gullible. To bring this about the decision was made to use modified religion without Christianity as a tool, as they knew they could convince enough people to follow them; which would allow them the best opportunity to guide the people in which ever direction they chose.

The cost of deciphering this purpose built deception that would be given, was deemed to be so valuable as to warrant a levy of ten percent of a persons earnings. The great prize for following this interpretation was heaven, absolutely guaranteed upon death if you kept up the ten percent payments. If you did not follow or pay, the most horrible things would dog you through out eternity.

With these decisions made and being applied, calling themselves a Sect instead of a tribe, then changing the tribe name to Pharisees. It was necessary to corrupt the original writings to suit their designed purpose so they took a lead from the Herodians and started their own interpretations of the five books of Moses. It was at this time they decided there would need to be a system for intimidation and so the words and meaning of heresy and blaspheme were bought into being. These two words were used to have Jesus crucified and have cost much stigma plus many millions of lives.

The whole idea behind this was to implement the form decided by the Pharisees (it must be my way) for they knew not anything of the spirit therefore could not represent the teaching of Jesus. It was the best way to relive government or individuals of their power, implement their own agenda and create the fear of hell.

As like breeds like and this tribe still retained their ferocious inbreed character they had no difficulty in enlisting the aid of the scribes (accountants) and lawyers for a reasonable fee. The Spartans, being the hit men of the day were employed to win any unwinable dispute with the guarantee of absolution for upholding the will of the Illusionary God of materialism, and save the Pharisees getting blood on their hands, therefore retaining their supposed respectability.

This did not mean the silencing of all opposition as the strongest part of their ploy was conflict in the mind of the people, and they knew the greatest lie tempered with the smallest bit of truth made it easer for them to advance their selfish greed and lust. The same as it still is 2150 years later

Before, during, and after the time of Solomon the woman folk were held in very high esteem, and when taken as a wife were looked upon as a great treasure, which indeed they were for that special ability they have as home makers, instilling peace and love to the whole family.

When the Pharisees started their take over, disruption was the first of their intentions to be implemented. The given word with a hand shake was no longer trusted. You knew that behind that smiling face with the reassuring words there was deliberate untruth. Hatred and mistrust were stealthily pushed to the limit destroying the common bond between people and replacing love for what has now become known as sex and egotism.

This started with the good time to be had with wine, woman, and song. and the reduction of the esteem in which women were held. This started what has been a very slow demise of the family. It has taken all these years to break down the bond of love within the family unit, but that bond is now returning with greater reward for any family participating even slightly in the ideal principle, for people have realized that enough is enough.

The attitude of the Pharisees to woman is clear from the knowledge that they changed their women almost as often as they changed their night attire. To give some idea of the disgusting depth their thought had taken them it was customary in the name of the illusion they called God for them to take a house away from a woman if there were no male occupant, and sell it to the highest bidder, pocketing the money.

This would preserve the woman from becoming a prostitutes. In fact it forced many into that profession for which the priesthood judgment was for the woman to be stoned to death. Sit quietly for a moment and feel that experience, ordained by some man who claimed to represent a forgiving and loving God.

From the time of the Pharisees forming into a sect until Jesus started his ministry; a period of about 150 years the Pharisees had made great changes in the truth of what Christianity was all about, and at the same time become financially wealthy, and spiritual paupers.

The Hebrew community also discussed and rewrote the five books of Moses. It was customary for all interpretations by any group to be mystical, with a truth of that time's academic, not spiritual understanding.

You must never forget it is not necessary for any to know how this system works, just do it, be it, be only in the present, presenting the highest ideal your conscience is capable of producing, then living it without fear and giving only love. This is what Jesus demonstrated for he knew people of the time would never understand the technological aspect because of the limited knowledge available and the scoff of the temple academic's.

The method the Pharisees employed after taking control of the temple and to create this wealth was that, in the inner Synagogues were large wooden casks into which followers of the Sect would place their ten percent offering. These casks were called the treasury. People placed in these treasuries all manner of things, such as clothing, material, food, money, etc. The amount of goods received were often surplus to the priesthood requirement so the temple traders business was to convert the excess into cash. This cash could then be loan at a high rate of interest on behalf of the priesthood to those who were having trouble creating their ten percent. This was the beginning of money lending even although it was illegal.

The temple traders were bonded to the priesthood with contracts drawn up and held in safe keeping by the priesthood. The traders paid the lawyer his excessive due as it was considered good permanent business with good profit. The lawyer collected money and interest on behalf of the priesthood plus his commission

It was against Roman law to loan or borrow money hence the temple trader to keep the hands of the Pharisees clean and as the priesthood held the contracts there was never any proof. All these things were happening without the knowledge of the people at large and any query for truth was met with positive denial. This eventually became the financial system which was born through corruption in contempt of the law.

Jesus had fear of no man so as he spoke the truth it was a pill the bigoted could not and still can not accept. It is just a lack of understanding the technology of the spiritual demonstrations undertaken by Jesus, and the lack of the language available at that time to explain this technology.

This lack was reinforced by those who would lose control if the technology was understood.

These are the miracles performed by Jesus and others with his knowledge. You still have access to them as you are also spirit, all you need do is to accept, hold fast, and implement the principal ideal of the universe through your expression of life.

Here is an example of truth faces bigot.

When Jesus accepted the invitation to dine with the bigoted Pharisee, he did not wash as was the custom of the Pharisees. As Jesus understood the thought of his host he said to him "Your sect are remarkably careful to keep everything clean that touches your food, least by eating it your body should be polluted; but you take no pains to clean your minds from the pollution of rapine, covetousness, and wickedness. You must surely be convinced that he who created the body formed also the soul; and can you imagine the Almighty who requires purity of the body because it is the work of his hands, will not also insist upon a greater purity of soul which is undoubtedly the far nobler part of human nature?

Instead, therefore, of that scrupulous solicitude of washing your hands when you sit down to meat, Ye should be careful to apply yourselves to the great duty of charity; (love) a duty that will render it impossible for any external thing to defile you, but will at all times be acceptable to your maker."

But the Pharisee, obstinate and perverse, withstood every means made use of by the benign Redeemer of mankind to conquer their prejudices, and bring them to the knowledge of truth.

Jesus continued, "Woe unto you, Pharisees! for you tithe mint and rue, in all manner of herbs, and pass over judgment and love of God ; these ought you to have done, and not to leave to others undone. Woe unto you, Pharisees! for you love the uppermost seat in the Synagogues, and greaten in the markets. Woe unto you, Scribes and Pharisees, hypocrites! for ye are as graves which appear not, and the men that walk over them are not aware of them." (they set traps of untruth by not divulging their true intention). Does that sound political or commercial?)

A certain lawyer, who sat at the table, thinking that this rebuke, though leveled principally against the Scribes and Pharisees, affected his order also, was displeased. But Jesus who never had any regard to the person of men, disposed his resentment and told him freely what he thought of their character.

"Woe unto you also, ye lawyers! for ye lade men with burdens grievous to be borne, and you yourselves touch not the burdens with one of your fingers."

Jesus also reproved the lawyers for filling the minds of the people with notions founded on wrong interpretations of Scripture, where by they were prejudice against the gospel; not being content with rejecting it themselves, but took care to hinder others from receiving it.

"Woe unto you, lawyers! for ye have taken away the key to knowledge; ye enter not in yourselves, and them that were entering in ye hindered."

The following comment is very important as it explains the results of misleading academic thinking verses the true spiritual application of the principle ideal of the universe

Speaking to all at the table and in ear-shot Jesus said "Woe unto you! for ye build the sepulchers of the prophets, and your fathers killed them. Truly ye bear witness that ye allow the deeds of your fathers; for they indeed killed them, and ye build their sepulchers. Therefore also said the wisdom of God, I will send them prophets and apostles, and some of them they shall slay and persecute; that the blood of all the prophets, which was shed from the foundation of the world, may be required of this generation; from the blood of Abel unto the blood of Zacharias, which perished between the alter and the temple; Verily I say unto you, it will be required of this generation." (In the area of no time and space we in 2005 are still part of that generation.)

It is very clear from the above that Jesus took the lawyers to task for enticing people into contracts which were no more than a burden which could not be shaken off and caused the contractor to commit all sorts of things to fulfil his contractual quota. It is also interesting to note that the lawyers were enticing people away from the truth of Christianity, or was this part of the obligation to obtain the Pharisees business and further their agenda.

The mood of the time suggests it was a soul selling for a chance to make money and gather material things, which still continues, Equally interesting and of even more importance is the rebuke given to all for condoning the building of the sepulcher of the prophets. It is proof that the originators the descendants and supporters of the anti Christian plot of the last 2000 years will be held responsible, for as you know Christians who spread Gods truth by word and action are still persecuted. Jesus was not the mister nice guy where the truth was concerned and right through the time of his ministry this was his style.

This is a re-sue-me of the different sects of the Jews, a people with whom Jesus was most intimately concerned, both as an elucidation of many circumstances, as well as verification of many things, foretold concerning the Messiah. Josephus reckons four principal sects among the Jews; namely the Pharisees, the Sadducees, called also the Herodians, the Essenians, and the Galileans. The Evangelists mention only the Pharisees and Sadducees. To the Evangelists the rise of the Pharisees was unknown.

The Pharisees indeed claim the celebrated Hillel for their founder, as he is, by some, supposed to have lived during the pontificate of Jonathan, about two hundred and fifty years before the birth of Christ; but others with more reason, suppose he was contemporary with the famous Someas, who lived about the time of Herod, "long before whom the sect of the Pharisees" was held in the greatest repute.

It is, therefore, probable, that they claim Hiller rather as an ornament than as the author of their sect. One of the most famous tenets of the Pharisees was, that of an oral tradition handed down from Moses, and to which they attributed the same Divine authority as to the sacred books. This being strenuously opposed by the Sadducees and Samaritans rendered these equally detested by them. But none more incurred their hatred than Jesus, who embraced every occasion of reproving them for the unjustifiable preference they gave this pretended tradition to the written word of God, and for condemning those as apostates worthy of death, who did not pay the same, or even a greater regard, to the former than to the later.

Another tenet they embraced, in opposition to the Sadducees, was, that of the existence of angels, the immortality of soul, the resurrection of the dead, and future rewards. But, with regard to the last, they excluded all who were notoriously wicked from having any share in the pleasures of eternity; supposing, that as soon as death had put a period to their existence, their souls were conveyed into everlasting punishment.

A third tenet was, that all things were subject to fate, or, as some expressed it, to the heavens, except the fear of God. (Then it was not easy to conceive what they meant by this) Josephus, indeed, will have it, that they designed to reconcile the fatality or predestination of the Essenians, with the free-will of the Sadducees,. If so, this is not the only absurdity, or even contradiction which they held; but a certain learned prelate seems to have proved, that they attributed all to fate, or to that chain of causes to which their Creator had subjected all things from the beginning; among which the influence of the heavenly bodies was considered as the principal.

This seems to be hinted at by St. James, in the beginning of his epistle to the new converts, where he explodes that pharisaical leaven by the most beautiful opposition of the immutability of God, the giver of all good, to the mutability of the planets, which, according to that notion, must necessarily vary their aspect from a malign to a benevolent one, and the contrary, even by their natural motion and change of position.

This tenet of the Pharisees was, therefore, a new source of dislike to the doctrines delivered by Jesus, as these affirm that MAN AND WOMAN INDIVIDUALLY, ARE THE AUTHORS of their own unbelief, disobedience, and obstinacy; and, consequently, answerable for that, and all the train of evils these vices draw after them That is to say, the people were taught by the Pharisees things which were not of the teaching of Jesus and held responsible for their actions by the Pharisees.

But the most distinguished character of the Pharisees, and that which rendered them most obnoxious to the just censure of Jesus, was, their supererogatory attachment to ceremonial law, their frequent washing, fasting, and praying; their giving of alms publicly, seeking for proselyte, scrupulous tithing, affected gravity of dress, gesture, and mortified looks; their building the tombs of the prophets, to tell the world that they were more righteous than their ancestors, who murdered them, though they were themselves plotting the death of one greater than all the prophets; (Jesus) their over scrupulous observance of the Sabbath, to the exclusion of works of greater charity, and many other of the like nature; while they were wholly negligent of the moral and eternal law of mercy and justice; of charity, humility, and the like, indispensable virtues.

The very best of them contented themselves with abstaining from the actual committing of any enormous act, while they indulged themselves in the most wicked thoughts and desires. Nay, some, more hardened in their vices, made no scruple not only of coveting but destroying poor widows' houses; of committing the vilest oppressions, injustices, and cruelties, and of encouraging these enormities in their followers, under the specious cloak of religion and sanctity.

Well, therefore, might the great Redeemer of mankind compare them to whited sepulchers, beautiful, indeed, without, but within full of rottenness and corruption. 'Where do we find that today'

It is clear from this that the fear of the illusionary God of those who would control the whole of the peoples of the world is still relevant today. How long does it take intelligent people to uncover the truth. Well, the grip and what we call the church and commerce is crumbling very fast as some reject the whole of religion and question the manner of commerce and finance, even when the truth of the teachings of Jesus are being pursued by an ever increasing number of people.

The last erroneous opinion we shall mention of the Pharisees, common, indeed, to all the other sects but more exactly conformable to their haughty, rapacious, and cruel temper, was, their expectation of a glorious, a conquering Messiah, who was to bring the whole world under the Jewish yoke; so that there was scarce an inhabitant of Jerusalem, however mean, that did not expect to be made a governor of some opulent province under that wonderful prince. "Is this thought still prevalent today."?

How unlikely was it, then, that the preaching of the humble Jesus, whose doctrine breathed nothing but humility, peace, sincerity, an universal love, and beneficence, contempt for the world, should ever be relished by that proud, that covetous, that hypocritical sect, or even by the rest of the people, while these, their academic teachers, so strenuously opposed it?

The sect of the Sadducees is said to have been founded by one Saddoc, a disciple of Antigonus of Socho. Their chief tenet was, that our serving God ought to be free either of from slavish fear of punishment, or from selfish hope of reward: that it should be disinterested, and flow only from the pure love and fear of the Supreme Being. They added, that God was the only immaterial being; in consequence of which they denied the existence of angels, or any spiritual substance, except the Almighty himself.

It is, therefore, no wonder, that the Sadducees should take every opportunity of opposing and ridiculing the doctrine of the resurrection.

Another of their tenets, equally opposite to the Pharisees, and to the doctrine of Christ, was, that man was constituted absolute master of all his actions, and stood in no need of any assistance to choose or act; for this reason they were always very severe in their sentences, when they sat as judges. They rejected all the pretended oral traditions of the Pharisees, admitting only the text of the sacred books, and preferred those of Moses to all the rest of the inspired writings. They are charged with some other erroneous tenets, by Josephus and the Talumdists; but those already mentioned are abundantly sufficient for the purpose.

The notion of a future life, universal judgment, eternal rewards and punishments to men, whom a contrary doctrine had long soothed into luxury, and an over fondness for temporal happiness, which they considered as the only reward for their obedience, must, of necessity, appear strange and frightful; and., as such, could not fail of meeting with the strongest opposition from the; especially if we add what Josephus observed, that they were in general men of the greatest quality and opulence, and, consequently, to apt to prefer the pleasures and grandeur of this life to those of another.

The sect of the Galileans, or Gaulonites, so called from Judas, the Galilean, or Gaulonite, appeared soon after the banishment of Archelaus, when his territories were made a Roman province, and the government given to Coponius. For the Jews considered this as an open attempt to reduce them to slavery, Judas (an academic not the betrayer) took advantage of their discontent; and to ripen them for a insurrection, Augustus furnished them with a plausible pretense, by issuing, about this time, an edict for surveying the whole province of Syria, and laying on it a proportionable tax. Judus, therefore, who was a man of common ambition, took occasion from this incident to display all his eloquence, in order to convince the Jews that such a submission was nothing less than base idolatry, and placing men on a level with the God of Jacob, who was the only Lord and Sovereign that could challenge THEIR obedience and subjection.

The party he drew after him became in a short time so considerable, that they threw everything into confusion, laid the foundation for those frightful consequences that ensured, and which did not end but with the destruction of Jerusalem.

The Sect known as the Essences were established about the same time with the main purpose of having a master come and show the truth of God. These are the people who wrote the dead sea scrolls, which modern scholars have ridiculed to keep the twist in the truth.

The Essenians, though not mentioned by the Evangelists, made a very considerable sect among the Jews; and are highly celebrated by Josephus, Philo, Pliny, and several Christian writers, both ancient and modern. It is impossible to trace their origin, or even the etymology of their name. This, however, is certain, that they were settled in Judea in the time of Jonathan, the brother and successor of Judus Maccabeus..

The Essenians distinguished themselves by their rules and manner of life into two classes; the laborious, and the contemplative. The former divided their time between prayer and labour, such as the exercise of some handicrafts, or the cultivation of some particular spot of ground, were they planted and sowed such roots, corn, etc. as served as their food.

The latter, between prayer, contemplation, and study. In this last they confined themselves to the sacred books and mortality, without troubling themselves with any other branch of philosophy.

Both the contemplative and laborious had their synagogues, their stated hours of prayer, and for reading and expounding the sacred books. The latter was always preformed by their elders, who were seated at the upper end of the synagogue, according to their seniority; while the younger, who were permitted to read the lessons, were placed at the lower. Their expositions were generally of the theological kind, in which they seemed to have excelled all their Jewish Brethren. But they paid the greatest regard to the five books of Moses. (the first five books of the Bible) and considered that lawgiver as the head of all the inspired penmen:

They even condemned to immediate death whoever spoke disrespectfully either of him or his writings. Upon this account they studied, read, and expounded more than all the rest, and seem to have drawn their religion chiefly from the Pentateuch. The doctrine and expositions of the elders were received with implicit faith, and in their practice they conformed with an entire submission to all their sect.

With respect to their faith, they believed the existence of angels, the immortality of the soul, and a future state of reward and punishments, like the Pharisees; but seem to have had no notion of the resurrection. They considered the souls of men as composed of the subtle ether, which immediately after separation from the body, or from the cage or prison, as they affected to call it, were adjudged to a place of endless happiness or misery: that the good took their flight over the ocean, to some warm or delightful regions prepared for them, while the wicked were conveyed to some cold and intemperate climates, where they were left to groan under an inexpressible weight of misery.

They were likewise entirely adverse to the Sadducean doctrine of free-will, attributing all to an eternal fatality, or chain of causes, little different from that of Spinosa. They were adverse to all kinds of oaths; affirming that mans life ought to be such, that he may be credited without them. "And the author said Ah Men"

The contemplative sort placed the excellency of their meditative life in raising their minds so far above the earth, as to be able to see from thence what was done in heaven: when they had attained this degree of excellency, they acquired the character of prophets. In their practice they excelled all the other sects in austerity.

If we may credit Philo, it was a fundamental maxim with them, upon their entrance into the contemplative life, to renounce the world, and to divide among their friends and relations their properties and estates. They never eat till after sun-set, and the best of their food was coarse bread, a little salt, and a few stomachic herbs. Their clothing was made of coarse wool, plain, but white.

They condemned all sorts of unction and perfumes, as being luxurious and effeminate. Their beds were hard, and their sleep short. Their superiors were generally chosen according to seniority, unless there started up among the brotherhood some more conspicuous for learning, piety, or prophetic spirit. Some of them, indeed, were so contemplative, that they never stirred out of their cell, or even looked out of their window, during the whole week, spending their time in reading the sacred books, and writing comments on them. On the Sabbath-day they repaired to their synagogues early in the morning, and continued there the whole day in prayers, singing of psalms, or expounding the sacred books.

Jesus, in the audience of a vast multitude, gave a charge to his disciples in general to beware of the leaven of the Pharisees, namely hypocrisy: by saying. "Beware ye of the leaven of the Pharisees, which is hypocrisy. For there is nothing covered that shall not be revealed:

Neither hid, that shall not be known. Therefore, whatsoever ye have spoken in darkness shall be heard in light; and that which ye have spoken in the ear, in the closet, shall be proclaimed upon the house tops," Luke 12- 1-3

He added "Fear not the malice of the human race; it can extend no farther than the destruction of the body; your soul may bid defiance to their impotent race. Better, an undaunted resolution in the performance of their duty, founded on a firm confidence in God.

Take care, therefore, that thy soul is so completely enlightened by the spirit, that the emanation of light be not in the least interrupted by any evil passion or affection; that all the faculties of the soul may be as much enlightened and assisted as the members of the body are by the bright shining light of the candle. (The light of the creator)

During the ministry of Jesus he moved among the people as much as possible teaching mainly in parables, as the language and understanding of the day did not allow for the lingual explanation of a technology which even in this supposedly technological age is still not understood.

This created the need to explain in two different ways. First was by the parables which was a simple story saturated with truth, even if not totally understood in the beginning. Faith was the necessary ingredient to be supplied by the listener of the parable, and those who demonstrated they had, and followed the faith unsparingly where those to whom all things were added, as they will be for those who follow the wisdom of the Christ

Second was by the demonstration of Jesus, what God is, how it was to be used, and how to gain the knowledge and ability to allow that God power to flow through you, and be accelerated by you for the good of all.

The parables are in the Bible for those who wish to know them, so I will deal with the demonstrations of the highest technology humanity is yet to understand, or better known as miracles.

It must be quite clearly understood that Jesus is NOT God the father, nor is he Divine Mind, and that he is as much God as you and me. That he did become, and is a CHRIST is absolutely undeniable. It is the edict of God the father that your future development is to develop the Christ of yourself for that is the spark of life, the soul, the spirit, the I AM, the real you, the whole I, total, complete, responsible, honest and in control of the self, for the benefit of all.

His message still is. you are spirit trapped in animal tissue by your own stubborn attitude to the things you will not understand, and that you do not want to use your perception to see.

In the world of understanding in which Jesus had and still has his being, his actions were common place and looked upon by him as demonstrations for those who were or would be his students. Because of the lack of understanding of the technology used by Jesus, we still call them miracles.

One of the basic necessities required by this system is to be of service to all, and that Jesus was to all who prevailed upon him.

His first public demonstration was of supply, when his mother Mary asked him to create the wine at a wedding reception. (how did Mary know Jesus had that ability?) He asked for the vessels to be filled with water and when this was done he turned it into wine.

Another demonstration of supply was with the bread and fishes. The multitude fed to capacity at this time was calculated at ten thousand people. It is little known that he also fed a like number of people who had followed him into the desert for a period of three days using the same ideal principle of the universe, the thing all people must eventually come to understand. Also little known is the fact that when he had fed the multitude, the amount of fish and loaves were greater than when he started.

Another reason people followed Jesus was that at all these mass congregations, those who had not lived within the law of the Divine ideal principle and become unwell, were blessed and made the "Whole I" again.

When Jesus raised the man from the dead he was still bound with the death cloths, covered with soil, and had the stench of a decomposing body about him. This service was carried out because this man, the brother of the woman asking the miracle was the only man in the house, and Jesus had no desire to see the Pharisees turn the woman onto the street.

Also when Jesus was on the cross he gave the charge of his mother to the ever beloved John so as she herself would not also be turned onto the street.

The Bible gives more detail of these and other demonstrations carried out by Jesus for those who wish to follow the truth of this story.

This world's greatest disaster happened when the man who through three trials in the courts of the Romans, was found innocent of all charges laid by the Pharisees, was crucified.

It was Pilate who told the accusers to behold the man. The Pharisees charge against Jesus was blasphemy in the eyes of the so called church, but by their cunning they managed to, and the fact that it was preordained, have Jesus crucified.

This was all part of the purpose of the coming of Jesus into the world, his life was preordained. He was born of woman, an immaculate conception, as it can be for any who follow the tenant of divine principle. He lived the physical life of an ordinary person, he was educated at and rejected the education of the mystery schools of the world. In deep contemplative thought he found his way to the truth, then proved it before starting his ministry through demonstrations and parables.

When Jesus eventually died on the cross it was the absolute end of mans endeavor to silence his message (for the Pharisees) for ever. They could commit no more physical atrocities against him. For the Pharisees it was done, it was finished, they were the victors, they still think the same in this day and age. Not surprisingly it was still only a part of the demonstration Jesus came to divulge and prove to the people.

In the tomb, the spirit of Jesus was so strong that it was able to repair, and return life to the body, thereby allowing Jesus to raise his frequency vibration to the point of being invisible to any whose vibrations were too low. When communicating with others he was obliged to raise their frequency vibration to do so.

This is the method he used to prove that life was continuous, and that you can only learn the secrets of life while you have life. If the learning is not done this time around, after justification in the astral dimension, reincarnation will see to it that you will have to go through all the trials and tribulations, lies and deceit, in general the hell which humanity has allowed to be created by the few, for the torment of the many unprepared to learn.

I can see the Pharisees jumping for joy as they see their two thousand year old deception still holds sway with the people, keeping them in the illusion created by the Pharisees for their own purpose.

To the absolute disgust of the Pharisees the demonstrations of Jesus and the word of God was spreading even faster than before. You must remember that what ever is stored above the eyebrows will eventually trickle out of the mouth, or through the sheer pressure of thought be put into action. It was then that the Pharisees used all their cunning in intimidating as many as possible into persecuting and bringing to the Pharisees for just? trial, those who became Christians.

It is not necessary for me to detail these things as they are well described in the Bible, although it makes one wonder at the number of people who would rather face death by stoning, or being torn to pieces by lions sooner than give up on the truth of the way of God as shown by Jesus.

The deciples of Jesus were not mercenary, and when Paul returned from two years in Rome and other countries he brought back and gave to the Pharisees a very large amount of treasure and valuables. There were twelve helpers in Paul's entourage handling the pack animals.

The hit men of the day were known as the Spartans. They were employed by the Pharisees to do the dirty work in return for a positive place in heaven. As they became accustomed to the dirty work they took it upon themselves to carry out any corrective measures that suited them. The situation became so untenable the Romans destroyed the Pharisees temple and banished the Jews from the Holy land, this included the Spartans.

It was during this banishment while slaves of the Babylonians that the Pharisees bought their way out of slavery and became goldsmiths.

Times were difficult for people to carry money about with them and the vaults of the goldsmith was the only safe keeping for the traveller. Into the care of the goldsmith anyone could deposit their gold, to which the law required the goldsmith to give a receipt. This receipt was only by law redeemable by the person making the deposit and loaning money or gold was punishable by death. But cunning and deceit for financial profit was born in that era and still reigns supposedly supreme today.

Here is an example of the goldsmiths art. A camel trader known to the goldsmith told the gold smith he knew a man who had travelled far to buy fifty camels, he also knew someone not too far away who had fifty camels going cheap, but he did not have the ready cash to carry the deal through, could the goldsmith help.

Realizing the financial implications, the goldsmith gave the trader a receipt for the purchase price of the camels, The deal was fifty fifty split less interest and a charge for book work, the money to be cash only. Buying for fifty and selling for one hundred and fifty left one hundred in gold.

With the gold on the counter the goldsmith did the divvy up. Interest on one hundred and fifty at twenty percent, after all he took all the risk, purchase price fifty, book keeping ten, and thirty each for their share.

That was the first cheque whereby the dealer received thirty for his work and the goldsmith received one hundred and twenty for the issue of a piece of paper which the trader imagined had value. The goldsmith still made good the fifty which represented the value of the receipt, and sat back in the glory of a deal well done with what he called earned reward, of seventy.

This was not only the first cheque it was also the beginning of banking as we know it. It also opened the idea for the goldsmith to loan money to the Babylonians at no interest for favours, such as law changes. This started financial corruption which eventually brought down the state of Babylon as it will with any state that is involved in corruption.

It was about this time the goldsmith realized if he increased interest it would inflate the value, and he set about to use this method to control the money supply. This system is still in use today although you will never get anyone to say so as inflating prices is dishonest.

After five hundred and twelve years of Christian persecution, the Pharisees realized they were missing out on huge amounts of tithes and not having control of many thousands of people.

To retain what they had and make a start on financially controlling the world, and making the world's population slaves to their system, they decided to start what would be called a Christian church. So the first Christian church was founded with all the high ideals of the Pharisees, the high alter, the burning of candles, the bobbing and bowing, the austere looks, the over done and expensive dress, the images by way of Jesus on the cross, the mother of Jesus, Mary, a language no one could understand, (Latin) the ten percent of your earnings, the total yielding of your will and seen actions, to put fear in the heart of men and woman that only any man who called himself priest was able to annul and give forgiveness, for a fee.

All the things Jesus rebuked the Pharisees for. In other words the organization was set up to use the word God only as a front, and pursue the Pharisees dream of world domination with the people as their slaves, the slaves were managed easier if they were given meager amounts of money and convinced that they were free providing they went to church.

This did not stop the persecution of the true Christian or those who saw through the deceit of that church, for the priests, bishops, and cardinals saw to it that those who did not pay their ten percent were ostracized by those who did, thus started the hatred on a personal basis between that church and other religions, in contradiction of the instruction from Jesus to love your neighbour.

How God loving is a leader or the followers who use a gun or explosives to proclaim to the world, you can have any opinion you wish as long as it is mine. Any man, group, or country is not following the tenants of the divine principle if he, it, or they imagine that my will be done instead of thy will shall be done.

Politicians, spin doctors, or corporations do not use divine principle, it seems to come naturally only to 'true statesmen' for they keep themselves their home, business, and country in a continuous expanding creative renewable peaceful environment by using the principle wisdom of Solomon "A win win situation for all.'

Very briefly I have put before you what the world's material thinking, grasping people have been like as leaders of the people. Therefore I can only say, the population at large must condone this leadership attitude in this day and age for it to still be the same; or has the mushroom been content with being kept in the dark and the fertilizer it has been fed.

You think this historical plan has vanished in the light of modern education? Every year since the dawn of history man has considered he was modern and well informed. Well here are a few pointers for you to consider.

Creditable education is only that which the student can remember of what the tutor gave out, 'right or wrong' so as to pass an examination. acceptable to those who control. In the courts of law the lawyer soon discredits Life's education as it has no piece of paper to back it up. WE compound this attitude of recognition of life's education as we seem not to remember the lessons of the past, and trudge along being guided and lead by those who give proof a plenty that they do not know. Friend, they have the piece of paper and are considered by others holding a piece of paper to be knowledgeable and responsible.

By and large all the takeovers, mergers etc, have been done by people employed by those whose interests are best served by the control on networks like food, distribution, rail, power, gas, petrol, water, sewerage, airways, shipping, roading, money, and making them monopolies even if under many names with very few owning the majority shares. In other words on today understanding of finance. if you control one or more companies, on paper it becomes more economical to merge, which of course gives you more power over those who make the decisions to make further mergers and takeovers. It must be reasonably clear to the average person what the mad rush takeover world wide is all about.

Redundant people who have been faithful to and defended the ideas of this system are the people who are paid exorbitant salaries and given responsibility to implement other technologies. The technologies being employed to close the gap and gain overall control are the poisoning of the soil and ground water through run off, then treat the water with other poisons under the pretence of purification.

The use of artificial fertilizer locks nutrition into the soil and kills beneficial soil bacteria and earth worms. Insecticide, herbicide, pesticide etc., impregnate the fruit and vegetables with homeopathic size doses of their poison, so depleting nutrition in plant and body, bringing the forerunner of disease through clogged liver, weakened or destroyed immune system, which enables you to accumulate massive medical accounts,

Stealing a life force so as to clone plant life, animals, and humans. 'Man cannot make the force which is life ' Genetic engineering of plant life through implanting foreign genes, virus, and all manner of other nasty things, terminator technology to stop seed from your plants germinating, patenting seed varieties which are thousands of years old, All under the horrendous lie that there is and will be even bigger food shortages on a world scale if these technologies are not implemented. "Fear is the enemy of truth and the opposite of love"

Even extending that lie to mean that these technologies are the only way the world population and wildlife will survive, and that they are so good for you, you can not live a healthy life without them,

To survive is one thing, although to live as man was created to live means he must not disturb the balance of the original sustenance that was created for his health and well being.

God in his creation gave man an organic garden of herbs to tend, so, as you seek an abundant healthy loving life you will be faithful to the creators hand work. God also gave you free choice and allows you to apply your own brand of intelligence God also gave you universal law which no matter what, you will never bend, and life to experience that law so as you will reach your ultimate potential. Not to be confused with the human selfish desire for commercial success and grandeur, and selfserving greed.

The above statement needs your further thought as man has been encouraged to accept it as only nature, over which he has further been encouraged to believe he has no control. Your duty to nature is to get off your butt, find how to, and tend the soil as it was intended by God. By not using artificial substance recommended by artificial Gods or the proponents of the death by decree of the artificial Gods. Although it is very far reaching with many helpful and harmful implications, your actions can be beneficial to yourself, your family, friends, and all life forms on and in the earth if you are applying it by the right-use-of. (right-eous-ness) universal law.

If your mind is not set in concrete and your interest is genuine with regard food and supposed food shortages, there is a book of absolute truth which will enlighten you beyond your wildest dreams. WORLD HUNGER; TWELVE MYTHS, SECOND EDITION U.S.$13 plus $4.50 shipping in U S Available from Food First _ institute Of Food And Development Policy, 308 60th street Oakland, C A 94618 U. S. A. phone (510) 654- 4400:

To give some idea of the books content, it goes like this. " In 1995 while at least 200 million Indians went hungry, India exported U.S$. 625 million worth of wheat and flour and U.S.$ 1.3 billion worth of rice, the two staples of the Indian diet. In addition, the American Association for the Advancement of Science found in a 1997 study that 78% of all malnourished children under five in the developed world live in countries with food surpluses..

A study of 15 countries found that per acre output on small farms can be four to five times higher than that on larger estate

Let us now consider our true selves, what we are, in which fashion are we connected to the cosmos and therefore universal forces and what is our power and path within this realm.

There is all sorts of understanding or misunderstanding regarding the physical form, the astral body, etheric body, the mental body the aura the ego, guides, soul, holy, spirit, the Lord, God, God the Father, Divine mind, universal law and a host of other things, designed by the unknowing for the purpose of expanding the conscience or deception, therefore the control over.

To clear some of the dust, which is fed to the mushrooms, we can first understand the relevant meaning of these words. Truth is light, food for the third eye, conscience expansion, deep internal understanding, so this will be the first step for you to see that light and if so desired follow it to your ultimate destination, for only total truth will add power to your being, word, and direction.

Divine Mind, is the Supreme Intelligence that created The Great Principle ideal Of The Universes, the ideal from which the word "GOD" and GOD itself was created. The IDEAL from which the law of constant ever lasting youth, constant physical and environmental sustainability and renewal is derived, the IDEAL from which it all springs, continues, and is held together, LOVE.

It is entirely up to you to make love unconditional in your daily lives with others, thereby opening the way for even higher ideals into expanding your consciousness so as you can express (implement) even higher ideals for self and all in your expression of life.

The energy underpinning this ideal principle is the atom, and the following is intended to create an understanding of the thing that makes all things one, as it is the thing from which all things seen and unseen are created. It is the only substance available to god to create

An Atom has a nuclei (curricular centre) around which small particles of energy orbit. It could in simplistic explanation be equated to our sun around which the planets and asteroids revolve. Atoms fill all space as solids, water, gases, colour or energy unseen, known by researchers as the aqueous mass.

As these minute particles of energy cycle around the nuclei, each within its own orbit, they create a three part energy which is electricity, magnetism, and frequency. This combination output of the collective atoms of the universe creates an energy which is commonly referred to as the Great White Light, or God the Father, as it is from this creative energy that all matter is made.

This is the matter, creating, and healing energy sort by all, and theorized by New Zealand scientist Brian St Clair Corcoran and confirmed at Stanford University where light was converted into matter in the form of an electron - anti - electron pair

Each part of this three part energy can be made more powerful than any other part so as each part can be used as a supposedly separate energy leading to the established theory that electricity, magnetism, and frequency are separate fields.

The things you need to have a fire are a combustible material, heat, and oxygen. Without any of the above you can not have a fire, and so it is with frequency, magnetism, and electricity.

The Atom being the building block of all the things we call matter (plus many things we are yet to gain understanding of) includes all animal vegetable and mineral and that means humans as well.

Because of this, the body responds to the frequencies encoded in the crystals we ingest in our food, liquids, and air we breath or the energy's we surround ourselves with through our own position in the environment or the process of thought which is a bio chemical electronic process.

This bio chemical electronic process is the self tuning unit which creates the resonation in the bio cosmic resonator. (the body)

The body takes in these frequencies, uses them for function, thought etc., then emits a frequency which can be measured for it's harmonic or inharmonic balance

Within the human body there is created four types of electricity, Analogue, Digital, A C and D C, so where you find electricity you also have magnetism and frequency. This is electronic or spirit or life force or God it is all the same thing.

This combination is the common factor in the holograph of life and what we call matter. Because of this it makes the animal and human a "Self Tuning Bio Cosmic Resonator" The acupuncture points and the endocrine system are the electrical interface and sub stations for the distribution and interchange of these energies.

This three part energy is what is known as spirit and what makes all living things an electronic unit. That is to say the cycles of the energy within the atom of the first cell is a constant ever lasting life force. It is indestructible and is the energy which leaves the body at the time we call death. It has intelligence beyond your imagination. It is the life energy which recounts and gives understanding of life experiences in the astral and accounts for that which you have or have not done in your previous life.

It has the intelligence to select from the astral recounts, the environment, country, physical, and social conditions in which you will return through reincarnation. This gives you the opportunity to experience and incorporate into your own character, attitude, and activity the highest ideal of universal law, or block it with that human element, greed, ego, me, mine, self. not to be confused with the 'I AM' for that, when you learn is the higher self in total control of your own physical.

The first cell is known as the central SUN of the body and it is this cell which draws other atoms known as solar SUNS to its self to create the physical unit. The central sun has the intelligence to create from the knowledge gained from the astral recounts, the type of structure that will best help you to gain the understanding that will give to you the truth, as truth sets you free

When free you will then be the SON who has returned to the Father's house.

All the above is Cyclotronic energy and can be measured by a skilled technician for balance or imbalance by the cyclotronic energy balancing unit.

It must be stated here that the current output of these electrical systems within the body is in the millionths and trillionths of an amp, so the use of high power or substance out of frequency balance with the body will blow fuses and bring on ill health.

You may or may not know that the Aura is a reflection in colour external to the body reflecting the internal health and or the mental attitude of the individual. Therefore like the ethereal bodies it is an electrical mirror of the state of the resonating frequency balance of the person in question.

Each component of the parts of the body has its own cycle and measuring those cycles reveals the inharmony of an organ or what ever, and can then be traced to the actual component that is not cycling in harmony with the other components.

The LORD is the law which governs this IDEAL principle of Divine Intelligence or is universal law. Right is might: use it and watch the twisting of the deceiver and his agents.

The light of the central sun of the universes is the source which sends out the penetrating unstoppable light bullets to collide with the groups of frequencies which we call matter. This does not destroy matter it only converts it to many more useable forms of frequency.

I believe this happened over the New Plymouth area of New Zealand at the turn of the century. Nostradamus suggested to look to the heavens at the turn of the century and right on cue it was filmed by at least two amateur movie photographers. The film showed a trail of smoke as if something had exploded. There were no particles found.

All frequency is energy. This is GOD the father, a process with intelligence, not a being, for in the shattering of the larger groups of frequencies the process has created the aqueous mass or the GOD stuff

GOD is the controller of these frequencies and when the son of the father calls for help in right use 'right-ous-ness' of the law, it is the father which directs the required frequencies through God who adds his power to it and sends it on its way to accomplish the task requested

The soul is a group of frequencies ordained at the beginning by divine principle to have dominion over the earth, (not owner ship by purchase or otherwise.) This is the thing we call the life force, it is the individual and indestructible, it will take up as many bodies as is necessary for its return to the house of the father, that will make you a light being with life ever lasting. The CHRIST of yourself.

Spirit is in modern language electronic. It gives you life to put into action, that which comes from the self tuning unit, or allow the highest ideal principle to flow through you for the benefit of self and all.

Which ever you choose will be displayed by you in your attitude, action, activities, principles and voice. This is something the spin doctor will never be able to hide for he highlights the truth by the lack of it in his presentation.

The various bodies are external sensory fields and the immediate vibratory extent of your central sun.

The ethereal, and auric bodies are the thought form of divine principle the mold into which the aqueous mass 'GOD STUFF' has been instilled to create the physical form. GOD made man HIS own image. "Man in his physical form has his own image and as a light being is in the image of the true God."

By fully expanding the central sun of your self with the energy particles of your own atomic structure moving at higher cyclic speeds in complete harmony with every other part of the atomic universe is obtained by stepping out of mans law, reach for and implement only universal law with the help of your self tuning unit then you will be the true god and the "Christ" of your self. "WHOLE I". The "SON" of god. The "I AM" Not the earth bound apprentice

The mental part of the physical body is the connection between higher and lower frequencies, "higher or lower thought and action"

The physical body is the power generator which creates electrical energy to drive the self tuning bio chemical unit for the use of Mind Body and Spirit, The whole I, The I AM. The subconscious mind is no more or less than a storehouse for past experience in this life time.

Holy is a distortion of the word "WHOLE I" It means that you are whole, total, complete. with all the attributes necessary to expand your consciousness through the right use of the principal ideal of the universe, even if you are unaware or try to deny it

Righteousness, Right use of, means the correct manner of the use of the great principle of the universe. The universal law. God's law.

Atone, At One, means to be as one by expanding your consciousness to the point of realizing and automatically acting with all things as one. As all of God's things are the product of harmonic frequency balance, how can we not be one?

Friend; I would like you to sit quietly for a while and give the following some consideration.

There is and has to be an intelligence out there, can you comprehend the magnitude of this intelligence. Can you comprehend the manufacture of the smallest spider, the grass, tree, earth, orb of the world, birds, air, cloud, rain, magnetic fields, or the countless cosmic forces so necessary to all forms of life.

What about the moon with enough energy to lift the point of earth facing it eight inches 200 mm, the rise and fall of the spring tides, the placing, maintenance and orbit of all things, this consideration could extend to the whole universe and beyond.

How inadequate is the word we understand as intelligence. How could that be explained two thousand years ago? Jesus new he could explain it, and he knew that it would not be understood, so he told all to have faith that his word was true. The word was with GOD and the word was actuality.

The world of GOD is the all embracing great rotundity field of harmonic frequency vibration within the adamantine and atomic structure, and when the word God is used with great depth of feeling and humility, it brings to the user the most powerful harmonic frequency vibration, and used in that context it will lift the frequency of the user, another step toward home. The word God used in repetition like a cracked record is reversing that effect.

You can understand now how with all the greed, theft, deceit, lies murder, the Pharisees, worshipers of the golden calf, the worshipers of the idol image of Jesus, Mary, and the Saints have all defeated their own purpose by keeping the word GOD out in front at all times without increasing their own or others conscious expansion of the true God.

It must also be remembered that those who had faith that there is a GOD and used their spirit to display it, played an even greater part in carrying the power of the law through at least these last two centuries. The true Christian is not involved in buildings or images, for he has found that the temple is the handy work of GOD, 'the body" and that within that temple he awaits your conscious discovery of him so as you can become one with him, and be your own true "SON of god"

He has given the temple over to your care and so as you can't say you do not know how, he has given you the little voice inside He placed all manner of herbs, grain, fruit and nuts in an organic garden on earth for your health, well being and conscious development

The Indians say there is no better hiding place for god than in man, For man has been educated that god is a being without, not the Christ within.

When true Christians were stoned to death they seemed not to notice the pain of the stones striking them, for the vision of the future being revealed to them overcome all the hate man could muster against them.

They went down praising GOD with the smile of forgiveness and love pursing through their bodies. This attitude was never understood by the instigators and perpetuators of this form of punishment for they imagined they had destroyed the all, They could not understand the working of the soul. Most still can not,

With an understanding of the above we can now take a closer look at the meaning of life. The female egg is a group of frequency and when invaded by the male counterpart it becomes a complete living expanding organism which attracts to its self the group of frequencies we call the soul. It now is a life force, it is the Christ within awaiting discovery by what will be this new individual. Brain washing will make it difficult for this new arrival. To help this soul during its pre birth development it will require the love, and devotion from both its co creator. The closeness of their being, the harmony of their spirit's their verbal contact and every thought will be registered in this new life force, it could also be a part of its future turmoil.

Another thing that can create problems for the new born is an inharmonic frequency of either or both co creators before and after conception. This is to say two people fortified with food from a source as close as was originally created in the beginning, better known in this day and age as organic. This high and nutritional food and water would best be continued from pre conception to death, Another source of problems for the new born and through out life is inharmonic things administered orally or injected

Even before birth and throughout life, this tiny soul is subjected to the conditioning we ourselves have and still are being subjected to for the sole purpose of total control over, in the name of profit, where as the expansion of consciousness into truth brings absolute freedom.

The spirit of this or any life is intermingled with the spirit of all life and life forms. The important thing to remember is that there is no matter, that all things are frequency. Now that this new soul has been nurtured and guided to the time of making its own decisions, it is confronted with two choices which will have the most far reaching consequences for its journey.

Either it is power position and money, wine woman and song or expanding consciousness into truth, internal wealth, and absolute freedom, " the integration of mind, body, and spirit bringing forth the WHOLE I the I AM"

WE looked at power, position, and financial wealth earlier in this book, so now lets get some more understanding on the things some wish to keep secret

You have decision making ability, a soul. Spirit is life in action.

Your own action develops experience and the learning is the knowledge gained from those experiences the total of which becomes wisdom.

Of course some people live and learn and others just live, so there are always exceptions to the rule, There are no exceptions to universal law. Those who do not learn do not open themselves for experience to self analysis.

I have referred to the foregoing message as a technology, and Jesus as the demonstrator of that technology, and from the following explanation it will be easy to understand how all the misunderstanding has come about.

The academic of yesterday were aware of as many facts as they are today, and for their world to exist, and them to retain their supposedly high standing in the community, even facts must stand to reason. Reason being the tool used to mask the facts, create myth and dogma and retain their position plus the status quo. (No shakers and movers allowed)

To give example of this in brief take the very air we rely on to have life itself.

In 1930 the percentage of oxygen in the air was 30 per cent. in the year 2000 there was 15 per cent oxygen in the air. The fast burning of fuels or explosions require more oxygen than the substance used helping to bring about the explosion. Reason says burn more fuel, have bigger and better explosions.

Trees convert carbon dioxide from the burning of petrol and diesel and other combustion into oxygen. Reason says denude the world of trees. Use them as fast and wastefully as possible, burn them to reduce the oxygen available and keep yourself warm. On no account must you use them to make a fuel which is environmentally friendly.

Mother nature makes a tree in her own time which we measure in years.

Man makes money by the millions on a printing press or more easily by the stroke of a pen in minutes or even seconds. Reason tells us we are short of money but the trees will be there for ever

Like commerce and finance, other systems such as pharmaceutical, medicine, science, for and against the well being of man, religion in all it's many misused and contaminated forms must have their own dogma which must be protected at all costs.

All know that truth will destroy dogma for the secret between truth and dogma is now being exposed.

The secret battle within the mind of mankind is like between two wolves. One is filled with evil, anger, envy, sorrow, regret, greed, arrogance, self pity, guilt, resentment, inferiority, lies, false pride, superiority, and ego.

The other carries good, joy, peace, love, hope, serenity, humility, kindness, benevolence, empathy, generosity, truth, compassion and faith.

'Which wolf wins the battle?' The one he feeds.

As man puts into action that which he thinks, he exposes to all which wolf he feeds.

If he feeds from the first group he is working from the intellect and ego through the mind.

If he feeds from the second group in a natural unthinking way he is working from his true self of love (the secret heart) giving it action through the spirit then the physical body, in which case the mind is not involved.

This is the difference between the secret heart which is the real you, and the person the mind portrays you as.

As said before these thoughts have power and the power which is generated by the first wolf could be called negative. It is the continuous passing through the mind of this kind of low frequency thought which creates the depression of the individual and eventually drags the one with those thoughts to the poor me, the chip on the shoulder, the world owes me something, I will do something (again negative) to show them, often in the name of God.

The truth is they do not have what it takes to face the reality of their illusion. They are hollow men.

The higher frequency generated by the thoughts of the second wolf are the ones which will raise your body vibrations allowing you to accomplish many greater things. Hereby lays the secret. When you have taken your body to the higher frequency you will then be able to let go and let God for you will then be able to lower your mind signal and live constantly in that signal to the theta level or the very low frequency range of out put. It is the lower range of frequency output which will do the work providing you are working from love. Surrender your mind to the universe. CAUTION if you are not having success reconsider your motives and if or not you are coming from love or ego

You have a car. It has a motor with pistons and valves going up and down. A crank, a cam shaft and wheels going round and round all enclosed in a comfort enhancing manner with a sleek good looking highly shining body. This equates to the human body.

Even when this body is fueled and lubricated it still lacks the ability to function, as the very life force the electronics which control and give life to this body are yet to be added. This equates to the human spirit as spirit is the human electronic system which in turn can be directed by the automatic out put of the friendly wolves. (LOVE)

It is truth of any subject that casts off the chains of other peoples belief systems that makes you free, depending on your realization of reality. This clearly demonstrates that you are not the car, but if you have the belief system that encourages you to think that you are the car, anyone will be able to take control of these functions and can be the motivating component, person, mental structure, or law. The motivating component, person, mental structure, or law will then be able to control this vehicle through the control of the belief system through the ego causing the mind to stand to reason.

This unit being under the control of those who would want to control others can enlist the cooperation of this unit, and through the food supplied to the first wolf have all in the same belief system recreating the past on a daily basis. (you are only a hired driver)

The center of the atom is the nuclei in which a certain combination of the adamantine particles thereof make the life force of the physical body. This is the love which you are and is the thing you feel when your heart goes out to someone or when you get that funny fuzzy feeling, that something special, overwhelming emotional warm happy feeling. The harmonic movement of these adamantine particles create the vibration of love. The vibration of the particles of love create electrical energy which is the beginning of electronics and is known as spirit. Spirit is also digital and controls the assembly of selected atomic particles and reduces that assembly of frequency into a lower vibration until it becomes visible matter. Therefore love is creating the tool for the construction of all things, as all things are of atomic structure

Step out of your belief system for a moment and feed your mind sumptuously only the feed of the second wolf. Where then would be the basis for you to proceed with the lies, deceit, dogma, hate, fear, hear say and division of the belief system which has controlled the academic reasoning of man, stripping him of his god given opportunity to follow in the steps of the one who came and demonstrated the truth.

How decadent is an administration where a person loves their country and fears their government, or a leader who stands before his people and says kill in the name of GOD, for is not the first commandment THOU SHALL NOT KILL

So what are you? You are not the car any more than you are your body.

The one thing we all have total control over is our attitude, and the fact that we are all an individual self tuning resonator, bound together and created by the electronic components of atomic structure, to create that which we call matter, or in this case a human being, with the capability of bio chemical electronic processing, (thinking) we have the choice of remaining in the human misery, or expanding our bio chemical electronic processing into the Christ Consciousness by the implementation of the God like attitudes.

IT was noted at that time by these cunning deceitful people setting up the system, that an army had only one supreme leader whose authority was not able to be challenged. It was also noted that this person of authority could be persuaded by persons outside of their control to implement ideas not necessarily in the best interest of those the authority figure was supposed to represent. These activities being fine tuned by controlled faulty reasoning over the last 2000 years have led to a bureaucratic army of system slaves representing various aspects of public concern.

One is called religion and is as divided as it is by the supposed knowledge of its many egotistic leaders wanting to be top dog, falsely representing good, joy, peace, love, hope, serenity, humility, kindness, benevolence, empathy, generosity, truth, compassion, and faith. These leaders expose themselves by condoning the activities of war which breaks the first commandment of thou shalt not kill, and plays into the hand of those who would divide and control.

Another is called government and is broken down into smaller over staffed units all reporting to a departmental head, who in turn tells the publicly elected representative of the whole community that which the departmental head thinks the representative wants to hear. This departmental head is considered by his associates to be above reproach, so is provided with enough financial reward to keep him as a yes man, thereby intimidating him or her with the loss of their integrity, power and position, or dispose of them by character assassination, (right or wrong) cultivated by those whose interests would best be served by maintenance of the tell tale signs of their true internal thoughts of evil, anger, sorrow, regret, greed, arrogance, self pity, guilt, resentment, inferiority, lies, false pride, superiority, ego and political correctness.

I would like to bring to your attention that every civilization throughout time has collapsed and disappeared because of the corruption, rottenness, lies, half truths, cheating and dog eat dog attitude. There is no reason why a major collapse will not happen again.

This is only a sample of the spirit of universal law and I leave it to the reader to find the connection of the self and its integration into the total of all that is.

To have an in depth understanding of the following, it will be helpful to understand the comments made by Albert Einstein when working on his theory of the unified field. "There is only mind and energy."

As there is no matter in the nuclei, there can be no matter in the atom, and if not in the atom [the building block of all things] there is only energy to create matter. This then means that the energy referred to by Einstein in the nuclei is a tight knit of frequency vibration.

Further, it is helpful to know that only things which have crystals within, will freeze. (Crystals themselves being created of frequency.) Water having the largest amount per unit of measure

Another thing necessary to be aware of is that crystals are and can be programmed, purified and reprogrammed.

If you look closely at a snow flake you will wonder at the number of crystals therein, Melt it and you have water with no visible crystals. This does not mean the crystals have gone or have no program.

The purpose of the crystal is to take up a programmed frequency from a source and transfer it to some other source,

When we have rain, hail, or snow it brings powerful healing frequencies to the earth to maintain the harmonic balance of the frequency combination we call life, all life. As snow on the mountain tops melts, it transfers the purity of its heaven gained frequency to the stone of the river, this then is used to purify or reprogramme any crystal which has been polluted by any other substance, only if the river is running with live water and is long enough to reprogramme all crystals.

In the food chain the purity of program within the crystals in the rain, transfer their frequency to purify the soil, and as the soil bacteria take up these frequencies and are ingested by plant life and other forms of soil life so the harmonic balance is passed along the food chain to the human and other animals.

The word pure meaning, that all frequencies held in the crystal at that time are of a harmonic balance and enhancing nature, with nature. (Absolutely no man made stuff,)

Therefore, if any pollutant is present or added to plant or soil the frequencies will be changed to that of the pollutant, and in most cases takes many years for nature to reprogramme. The word pollutant means any frequency or group of frequencies that are not in harmony with the planet or life forms on the planet, These are the killers of any life form.

We talk about minerals, vitamins, etc. and the relatively new photo-chemicals in plants. Minerals and vitamins if derived of a natural organic source are no more than a group of beneficial or harmonic frequencies stored within the crystalline structure of the liquids therein. Photo chemicals are the chemical frequencies within the plant and if obtained from fresh organic sources are proving to be of the purist most beneficial and harmonic frequencies for all forms of life, even in the compost heap.

The composting process is another way for frequency transfer particularly through the worm activity. Beneficial bacterial can live in an accelerated harmonic atomic environment as it is also spirit.

All life forms are only a harmonic group of cells shaped by the etheric body in the manner of that particular animal,

Each cell is like a city of frequency vibration although there are many different combinations of frequency that go to make up the various body parts.

The things we ingest exchange frequencies with our cells as the digested food passes through the system, So if you put in good fresh and a wide variety of organic food then health must follow as different foods contain various balances of frequencies.

The inharmonic frequencies transferred from the body to the passing crystal contain the toxins which are disposed of in the eliminations, Even these toxic crystals can be deprogrammed and reprogrammed. The composting toilet is one example. This is called the wondrous works of nature, or bacterial life in action, spirit.

Homeopathic remedies are just a transfer of frequency from one source to the crystal, thereby supplying frequencies which have not been ingested from the food chain

As the body is a group of cells all vibrating with frequency, exchanging frequencies within, it makes the body a self tuning bio chemical resonator, This coupled with thought which is a bio chemical electronic process, makes it possible for a correctly tuned mind to also program crystals and exchange frequencies with other outside resonators.

A quartz crystal well tuned to an individual will by action give a positive or negative response to a thought projected to that crystal.

The power of thought is the self tuning unit and the manipulator of energy.

Again remember that any and everything that lives is also part of the field of frequency vibration, even things which are in a state of rot or decomposition are still part of the same field.

Drugs, poison sprays. artificial fertilizers, alcohol, artificial sweeteners etc. quickly retard the self tuning ability of the self, causing breakdown

Over processed food, leaching of utensils, containers, plastic, etc. reprogramme the frequencies of the contents and has the same effect but does it slower.

This is only a brief description of the activities of an extremely large field of one hundred and eighty six billion frequencies. How mind boggling becomes the combinations available to the properly tuned resonator?

This has shown the life of the water in action has had an effect on many things, for had not the water been in action none of these things would have been possible. It is so of all life forms that when the life is in action the consequences are far reaching and have an effect on frequencies "life forms" far beyond our present understanding, like shaking a bottle of drinking water for five seconds will bring the water back to life.

Even the inert things like the stones in the river are still life forms as they are also made of frequency

Good hunting if it sparks the quest for an in depth understanding of life, universal law, how to use it, which of course will bring to you, the truth wherein you will find freedom.

Just sit quietly and think for a while about the following. The planet we live on, the various life forms that have inhabited it throughout its entire history, the timing and the growth of the food requirements as the very life blood of their existence, the water, the air, the resources, and resourcefulness of mother nature.

Did man create the smallest spider, the largest whale, or anything in between, no; they like all things are beyond the scope of even this generation. Man uses and in a colossal way abuses, for selfish reasons, all the things he cannot create. He fails dismally with his manic thinking and throw away attitude to even attempt to fit into Mother Nature's prime principle of continuous, renewable, sustainable everything.

Over the years he has moved further away from his own personal development, of discovering who he is, and the potential he has when he finds himself within, for it is truth which gives him freedom from illusionary belief systems.

These creations have all been put there by an intelligence far greater than man can possibly imagine, or even put into words for the purpose of explanation. The only word which could be understood by some would be DIVINE MIND in action implementing Divine Principle of creative, harmonic, continuous, renewable, sustainable everything. (and behold I make all things new)

Just think of the intelligence and knowledge required to start the building of a planet, or a galaxy, or even a universe. Now give consideration to the creation of life which to be correct must include all plants, grasses, insects, animals, fish, water borne life forms. the means of nutrition and nurturing of all these things. What about the complexity of the human body with all its bits and pieces, joints, blood flow, nutritional requirements and the thing which gives it life, their ills and health problems or their well being.

It is also known by a very small number of people that the human has not yet found their maximum ability to understand itself, for there is a greater depth of being yet to be discovered which exists inside the self.

This self off discovery so much talked about and entrenched with dogma, as God and religion, is the thing we commonly refer to as spirit, but it is a whole lot more with a very far reaching effect for the development and well being of the human and the very planet itself than that.

It is very much misunderstood, unknown. kept secret, used as commerce, murder committed in it's name, even denied with some fighting to the extent of nations at war over it, when all propound we are God's children with the virtues of peace and goodwill at our disposal through the spirit and the implementation of self in God like expression in our principles of daily living.

The people of the time of Jesus did see and on some occasions were involved in the mysteries brought about by His demonstrations, and two thousand years later those demonstrations are still not understood. Found not guilty after three trials, greed, fearing the loss of control, corruption of the truth and lack of understanding was the cause of this innocent and honest mans supposed death

Jesus also spoke of the wonders of the fathers kingdom assuring all it was available to them if they would live life as he did. "Non shall enter the kingdom of heaven but by me" He knew it was of no value to explain the technology behind his demonstrations as there was no scientific knowledge, of the kind necessary among the people of the time. His recommendation was to have the necessary unwavering faith in his word and pray to the Father for guidance,

Having said that it is now time to put the whole thing into as simple technological terms as possible for the understanding of the not so technically minded.

In the beginning there was the word, and the word was with God, and that word which was with God explained all that was, and still is, and is the word ACTUALITY. (in my dictionary that says realism; fact; reality; a thing which is in full existence.)

That of course was before creation had begun. What was this thing that was already in full existence? It was and still is a harmonic field of particles each particle in it's own orbit around a central field. Therefore the only resource available to make all the things that were created, and the things that have since, and are yet to be created is the atom. The scientific discovery of everything having an atomic number substantiates this point.

Without man's involvement the massive uncountable number of atoms which fill all space are in an harmonic balance, and are there and maintained by the Ideal Principle for the creation of everything and its returning to its original form. Thereby we have God the Father's gift to humanity of the continuous, sustainable, renewable, everything.

It may take many centuries for nature to recycle some things, although nature has the time, the question is, do you, before you are aware of the total irreversible man induced destruction occurring?

The life force is love, and you can only be true to yourself when you display yourself as love. It is something that the human with all their reasoning has no possibility of changing. This is the part of you which is God, and it is in this image we are created. It is the indestructible, incorruptible, higher intelligence of the self, the true self which will only be uncovered by the discovery of self within the temple, (the secret heart, the center of the life force, the God which awaits your discovery, your self as a Christ of the future which only you can develop)

Go within and there I shall meet you. Knock and you may enter, Seek and you will find.

This is the so called energy that is seen by some leaving a body as the life force withdraws, leaving an inert physical accumulation of spiritually developed cells which in time will return to the elements that spirit brought together in the hope that the physical would find its true self.

The center of the atom is the nuclei in which a certain combination of the adamantine particles thereof make the life force of the physical body. This is the love which you are and is the thing you feel when your heart goes out to some one or when you get that funny fuzzy feeling, that something special, overwhelming emotional warm happy feeling. The harmonic movement of these adamantine particles create the vibration of love. The vibration of the particles of love create electrical energy which is the beginning of electronics and is known as spirit. Spirit is also digital and controls the assembly of selected atomic particles and reduces that assembly of frequency into a lower vibration until it becomes visible matter. Therefore love is creating the tool for the construction of all things, as all things are of atomic structure

These two parts the adamantine particles and spirit together make the unit we know as the soul, the ultra ego, or the super ego (the trinity) often misguidedly called the spirit. As spirit is the worker, the life force, (the God given spark of life) directs its activities so as to create through atomic manipulation and the lowering of the vibration of that created, any object held in the thought pattern of the life force. This then is spirit in action and as it is directed by the life force it creates the equation that 'life is spirit in action' Then by that spiritual action embedded in humanity, progress in material well being is made.

Being unable to create in the above manner, humanity must create because of its less spiritual thought pattern and egotistic need to control and approach to creation, use already manipulated atomic structure lowered in vibration (God given resources) through their physical activity.

It must not be imagined that material progress is in any way the way to the inner self, it is only an easer means for the commercially selected few of keeping the physical available to accommodate the life force. Even faced with manmade financial poverty, hunger, starvation, and genocide the physical strives to stay available.

The universe and all it contains has been created by this method being used by Divine Mind through the use of principles of universal law as laid down by Divine Mind. The principles being known as God the Father as they are the principles by which all has come into creation. Therefore the principles of creation are the things we would be wise to follow and pay our deepest respect, gratitude and homage to, for they are principles of harmonic, continuous, sustainable, renewable, everything. (and behold I make all things new)

These principles and the implementation of have been entrusted to the human race. Have you personally abused that trust or helped others to do so. They are the only principles which will sustain the planet earth and therefore its inhabitance as they are the reality that is, it is God's creation not made by man, absolutely not the spin doctors reality that some human or humans has control over or is supposedly for the benefit of the under privileged.

Automatically it makes man a creative being for what other creature has the ability to design, build and fly an airplane, or any of the other things the industrial world has produced.

The giving out of the life force (God's gift of life) LOVE should not be interpreted as the sexual act, nor any act of self gratification. Only pure love given to all with no thought of receiving will be of spiritual benefit to the giver.

When in birth the first cell is fertilized a life force (adamantine particles of the nuclei) takes up residence. Its spirit then starts to draw to that cell the God stuff or part of the actuality or the atom (they are all the same) in the manner as set by the agenda of the soul. This life force has also taken into account the spiritual requirements of the unborn infant and any past karma it has to reconcile. It therefore also selects the parents who it knows are best suited to give the newborn and the parents every opportunity in reconciling their own past karma.

At the time the life force takes up residence it also relieves the parents spiritual energy from the cell and replaces it with the life force's own spirit. From that time on the cell is an entity in it's own right and starts, with the nurturing help of the mother, the process of drawing further atomic structure (and nutrition) required to build the other cells each an exact replica of the first cell which will complete the cycle of creative divine principle.

This life force being the unconditional love it is has now created a temple (body) with the God stuff in which to dwell. It is now known as a human and has been created in accordance with universal law as have all people irrespective of colour, creed, religion or nationality. How can we not all be one?

As all physical beings are created in the same way with the same ACTUALITY within the DIVINE IDEAL PRINCIPLE OF CREATION they have followed the method of the creation of the first man. God the Father created the son and that son is commonly called God so as all humanity is created in the likeness of the son we are all god although amazingly few have come to realize the connection. It is commonly known that the best place to hide anything is right under the nose.

It is written that God created man in his own image and that God is a spirit.

It is a more correct interpretation to say God is spirit and created species man his (not in) own image, not that of a wild dog, monkey, cow, horse, bull, ram, or mule even if the species likes to act like one. Your choice.

God's most noble creation has not been hidden except by mans incorrect interpretation as we are all walking God's or God man, all that must be done is to realize the truth of God and stand forward portraying yourself with an inwardly based character and an outward display of respect, honesty, integrity, compassion, benevolence, forgiveness and love seeing only the potential of the same in all others. If love and respect is not given how can you expect it to be returned.

Now it is known how humanity is created, we must now try to understand how it all works.

As the entire physical form is of atomic structure and the movement of the atoms create electricity, it makes the physical form an electronic unit which as a result will have an ability to put out electronic frequencies causing vibrations. The common terms used to describe the physical form is a word known as bio, and the other word used to create the physical form is cosmic.

With the soul having the spiritual agenda as its main objective to guide (the little voice inside) but not override the physical to the higher understanding or conscience expansion of its true and rightful place with full dominion over the universe, by holding in mind and enacting only highest God like divine ideal principle, and the physical electronic unit putting out and receiving electronic communications which is the thought out put of all other electronic units together with the egotistic application of the me, makes man a self tuning bio cosmic resonator.

Then with his own thought pattern generated within his own egotistic free will, and the belief of others that have been taken as a truth, is that which governs the out put of his electronic presentation (how he presents himself) to the outside world through the physical, is the system known as thought which is a bio chemical electronic process.

The learning or the expansion of the conscience starts at the moment of conception as the vibrations sent out by other self tuning bio cosmic resonators are registered in the memory of other resonators who pick up the communication. As the new resonator is at this time connected to a more powerful supply of spiritual energy (the mother) that supply becomes the main point of incoming communication. Therefore the relationship of the parents and those in close association are the foundation upon which the new life will start its own base egotistic self will thinking. It will not effect the agenda of the soul as that will be worked out when the new life takes on the decision making process. The important thing is to equip the new life with the best tool possible to make conscience expansion a natural part of being.

For ease of reference I will from here on refer to a self tuning bio cosmic resonator as a human being, and a bio chemical electronic process as thought.

Words and thoughts are things and are very much more important than most everyone imagines. So as not to go against the agenda of the soul the words are best to be a true verbal presentation of that which is held in the thoughts, or better still from the very core of your being for that is the real you. Take care for this is where you can expose your very core intention, for the word must match the action,

Thoughts are more powerful than words for words must be formulated by the electronics of the brain, it is therefore a spiritual activity which is compounded by the action undertaken. As thought is an electronic process it creates frequencies on the same level as the thought, (the self tuning) and those frequencies having come through the spiritual connection are the frequencies that control the atomic structure.

When you bake some scones you use similar ingredients as you would to make a cake or other types of baking, it is purely a matter of material combination according to the result you require. When you play music you have a number of octaves available, although the thirteen notes and half notes are the ones in which all the music is made. By using any combination of the above think of the variation and number of songs and music there are.

The field of atomic energy has many more particles so the various combinations are virtually unlimited. The baker manipulates the ingredients to make a cake, the musician manipulates the instrument to make music. what, how, and who manipulates the atom to create solid material objects?

The actuality or the atom (they are both the same thing) is the substance which is the raw material for the production of you, me, the planet and all other things in it, on it, or around it. As its particles are always in movement it interpenetrates all space and all other substance.

It is all other substance, and the particles never stop their movement ever. It is the God stuff, the creative energy, the all abundant everything waiting to be lowered in octave to become a material object.

The more solid the material the lower the vibration, the frequency out put of the atomic particle combination.

The continuous indestructible life force is given by God the Father, this is in his likeness. It is ever lasting and cannot be taken from you. Your body can be separated from it although in truth it goes on forever.

This is known as the central sun of the human, and is known in universal understanding as the SON OF GOD. (the meaning of the words SON OF GOD is IN UNION WITH THE FATHER) This is the real you (God man) around which the physical is build, like it or not, deny it if you wish, the belief or disbelief is for your assessment, it will not alter the fact of what is, or the actuality. (the real, the reality the fact, that which is and has always been in full existence) you are part of it, your being is proof of it

Love is the SON of God and is the central SUN (atom) of the physical. This unit draws atoms as the building blocks of the physical. The particles of these atoms revolve around their own nuclei and the total number of atoms revolve around the central sun.

Most of the people of the world are just sitting on their backsides hoping the Christ will come and take them up into heaven. I tell you friend your bum is going to get very sore while you wait. It just can not happen that way. Like it or not you are on a journey as is every person on earth, irrespective of any notions you may have, any other persons belief you may believe, any ideas of promised people, selected religion, or grid repairers and light workers.

You will not be lifted up. To fulfill the right use ness of the law you must make your own way.

To create heaven here on earth (that is where it is going to be, not some pie in the sky promise of someone who expects ten percent of your earning) you must get motivated, think only from the core of your being the God like thoughts, action only those thoughts, be the thoughts of God, at all times talk to God the Father direct, ask him for guidance, he is your real Father, your truest friend, your guiding light, he will be delighted that you have started your journey home to his house and his many mansions, be true to him and he will be true to you, when you work with principle, principle will work with you, show to all the love that you are, for that is your very core.

Every thought that has ever been thought on this planet and every word spoken has a frequency. This sets up the field of vibration. These frequencies bounce between the ionosphere and earth continuously, these are the records, here is stored all that which you will have to answer for. It is also known as the pool of knowledge as it contains all thought, spoken word, and every action, as every action requires thought.

The bad thoughts are there as are the good or God thoughts. It is from this pool of knowledge you can receive information through your self-tuning unit. (your brain) You can tune in to any frequency your free will dictates, whether its from the central Sun the Son of God, or any other outside source or other persons belief system.

The higher the ideal of the thought presented in action the higher will be your vibration. The opposite follows any bad thought or action to the point of reducing the self to wild animal status. Which ever path of thought you follow remember you and only you are responsible, Certainly the thought of your ego will be the first one you may like to follow although a moments delay before word or action allows the small voice in side time to render its opinion, then the action or word is up to you.

Implementing the highest ideals is how progress is made in spiritual advancement for it is a direct implementation of the soul's agenda,

The forgoing pages explain how it is that you are an electronic unit with all the potential of putting out good or bad frequencies. It also explains what happens when you continuously put out low vibes.

Another point to consider is the I AM. There is much misunderstanding about the saying of "I AM THAT I AM". When you vibrate at the same rate as spirit you are all things so it matters not what you want to be or do it is instantly available and so you can say I AM THAT I AM

Your interest could be in any subject or objective pursuit, it matters not as it all take thought, action, and dedication to get a worthwhile result. For the purpose of this explanation we could look at music.

You start by wanting to play a piano. This has established your requirement or the direction you intend to direct your thought and activity. It also engages your self tuning to that entire field of music, thereby opening the field of knowledge for your conscious connection. Study and practice will of course show the depth of commitment you have to your own ideal principle of music and add strength to your conscious connection with the pool of knowledge. Sit quietly and listen to the silence whenever the going gets tough, ask and be patient then you will receive.

We must suppose you are looking for perfection in this or any other endeavor as you are wise to seek perfection in all your undertakings, as perfection is a requirement of the spirit.

The wise will only entertain and implement the highest IDEAL for the spiritual development, for when you stop using your ego and work with spirit, spirit will be delighted to work with you.

As your musical ability develops your understanding of the whole world of music opens for your perusal and greater depth can be applied until you can in all honesty say I am a musician. You are now, and are also recognized as a musician. Truly you can now say I AM THAT (a musician) I AM, the that meaning musician.

This is how through desire, commitment, and dedication, perfection is brought into the life of any who have the will, determination, and integrity to accept and carry through to any ideal.

Your own path into the higher rate of vibration and conscious expansion can be achieved through the same method of desire, commitment, and dedication. If the desire is for the development of spirit the rewards will be an unlimited everything or anything, for you will then be not in but be the same frequency as spirit. This will be for you the second coming of the Christ as the Christ of your self, this is your true destination, you will be home in the Father's house then all things will be added unto you and greater things you will be able to do.

All of this can only be achieved while you are living a life on earth, then when you have gained this status you will have life everlasting and live wherever you like.

Should this be your desire, go into your own conscious as there are extremely few who would be able to help you on your journey. It is YOUR journey and in the end only you can travel and discover the true inner secrets of who and what you are.

You have discovered how the water was changed to wine, how the three loaves of bread and the five fish fed around ten thousand people, how Jesus fed a like number for three days whilst in the desert, how the water was stilled etc, etc. All these things you will do and more.

The other side of the I am that I am is the ME. this can be equated to how we live in this day and age, our own lack of knowledge and the determination to remain that way, The thought proven by our action of the way we behave. The resistance to finding the truth of who we really are and the latent ability which could be ours. The egotistic thinking, the use of meaningless words, the promises not kept or told with intent to deceive, the lust for self gratification, the insistence of self will at anybody's expense, the use of human might to install a particular style of rule over other countries, people, or person etc,.

The total of the above electronic technology opens the way for man in his present state to explore many avenues of this system. One such field may be the sound frequency of what is known as the audio life stream. It is a range of sound not too far above the range of human hearing, it is the very top of musical expression, it is totally harmonic and because of this the vibrations have an up lifting effect. The shifting of this sound to the human hearing range would produce an overall up lifting of all humanity for it is the music created by the movement of the collective atomic particles, commonly known as the Angels choir.

Another plus for humanity would be the use of the electrical energy produced by the movement of the uncountable number of atoms throughout the universe. These atoms and the electrical field created by the planet earth revolving within the ionized belts around earth, with their electrical discharge as lightening into the earth plane, gives a voltage of four hundred volts at shoulder height right around the world. This voltage can not be used until a method using electronics can be found to move the amperage, then the amount of voltage will be available according to the amperage required.

Units of this type could be as small or as large as required such as to drive a car, boat, ship, train, or power a house, factory, or even a city, all being independent units without the need of huge generating and distribution costs.

As mans thought occupies the extremely low frequency range, using the information herein could make it possible to electronically record onto tape or CD with colour and sound any event which have taken place during the course of our history complete with a panoramic view of the area where the incident took place. As thoughts are a progression of one another the frequency may have to be divided into minor points to obtain a contact for that precise thought pattern. Then as those patterns are contacted the pattern will give it the flow for acceptable recording results.

You can see by the above that mans thinking creates an extremely low frequency, a frequency close to disease and death, thus a very low life style. Lack of truth, other peoples opinions, spin doctors badly bent suggestions, the lure of material possessions, the lack of leader-ship in government or its departments, the world wide corruption, lies, and deceit all keep humanity at that low level.

This planet is struggling under the weight of ten thousand years of cultivating and harvesting the seed of disaster. That seed liberally fertilized with heavy loads of fear is the theory of it stands to reason, when seldom does reason stand to fact. Remember the fact is the reality, that which has always been in existence.

There has always been a choice but the mister nice guy back down approach has allowed the continuous reaping and harvesting of the bullshit fed mushrooms.

If you want a better world to live in, you, and I mean you, because each individual has a part to play and that part will have an effect on every other person on earth, as we are in fact all part of one. You will have to get off your backside and at least make a start to demand from all others and implement honesty, accountability and integrity with higher SPIRITUAL IDEALS for all humanity, and stop kidding yourself that some day you will be lifted up to heaven by a God who has no respect for the person of humanity, although the respect for the spirit of the person has Gods unlimited continuous out poring of love.

Like it or lump it, it is your journey and is only available while you have life. that is why if you do not raise your level of frequency this time around, reincarnation will give you another chance

Just a last little peep. This is the reality, the actuality, the real, the fact, that which is in full existence, it is undeniable, it is unchangeable, it can not be anything other than God, and you being in union are the son of God, God man or God woman, with the choice of becoming the Christ of yourself.

The time frame for the transition from that which you think you are to that of spirit and Christ hood is a mater for the individual, for if you open your ego to make that transition it will take many life times.

Then again if leaving what you call the physical world out, you open your consciousness to take in the all embracing completeness and fullness of the WORD God gave, "ACTUALITY, for it does in actual fact mean all that is and will ever be is in existence at the present time and has always and will always be in existence" and bring that into full realization knowing it to be the source of supply of all that you could wish, and have it materialized right before your eyes, is a matter of how long it takes for that reality to be completely assimilated by the mind and enacted by the physical in the total enfoldment of unconditional love.

It is no easy journey because of the misconceptions which abound all around us, and the method of spiteful personal ridicule used to deny truth, cover ignorance and for the liars to use the gullible to retain the corruption and the status quo It will take all the guts, grit, and integrity you can muster, and on accomplishment will prove to be worth the almost eternal grind.

A good place to start the journey is to reinvent and enact the loneliest words in the language "honesty, responsibility and accountability" giving them there true meaning within the meaning of the word actuality and implement it in all your ideals, actions, thinking, and sayings.

Pay little attention to the person of humanity, although to be of service to the spirit of the person will be to the advantage of all.

Allow others to stand on their own feet.

Recognize and implement freedom for you can in fact do what ever you like provided you are prepared to accept responsibility. It is when in your consciousness you allow the other people to do the same you have given freedom.

As you give the other people the freedom you consider yourself to have, you automatically release yourself from your own thought pattern of them and you are free.

Total complete bone shaking unnerving honesty and freedom are the two ingredients that go to make up democracy and without both being implemented within the true meaning of the words, democracy is only the myth of the spin doctors interpretation.

HEAVEN IS HERE ON EARTH, DO YOU HAVE WHAT IT TAKES
TO BRING IT ABOUT

The time has come on this planet for the supposed intelligence of the inhabitance to start understanding and implementing the reason for human existence. To do that it seems necessary to firstly get some facts correct.

Firstly it is well to understand that it has been the accepted attitude that anyone and everyone can and do, as they like without any one having any idea of the reaction of their thoughts, words, or activities. This has created the saying, thought, and the demand of MY will, will be done, and so we have the worlds people against one another to the point of other peoples life is of no value, it is my will that is important. But to whom?

The benefits of supporting the principle of the universe can not adequately be explained with the spoken word, it has not been well enough understood therefore it has not been able to be explained. Hence we have the dogma leading to ignorance through other peoples misguided opinion and belief systems.

This will change as soon as each, all, or any individual starts unreservedly supporting the principle, for it will be that THY WILL BE DONE ON EARTH AS IT IS IN HEAVEN. All of principle is all of heaven, the implementation into the individuals prime ideal of thought, and activity of the ideal principle of the universe is the path

The first requirement is to let go of that which has governed your reasoning (other peoples thoughts, the status quo, the ego, the written word, the supposedly scientific understanding, the established order of dogma held by those who would like to control, the hype, the flag waving, the basic profit reason for commerce, the basic reason for accepting other peoples expectations of you.)

These next lots will if it is done correctly can be done in an instant or can occupy many years of your time, although it will free you from the clutches of those who use you as the puppet you have been for all this time.

To verify that within you own experience, have you received the benefit from the last election promises. Is your health system delivering that which it said it would? Are the claimed medical breakthroughs ever implemented? Are the leaders of our nations God fearing enough to follow the teachings of love, peace, goodwill? Do they ensure an even distribution of Gods creation cultivated by man for all the people of the world? Do religious orders judge others to justify their own existence? Do you judge to justify your own? What in depth understanding have those responsible for the Divine Principle Of God Technology passed on? Do they know the Principle? Above all do they implement the Principle? With your hand on the Bible you swear to tell the truth, is judgment made on God's law or man's law?

I trust from the above you will be able to make the necessary divisions for your own deeper understanding of the world that is, and which it could be.

That's looking at the world, what about the people of the planet. It is only by honesty within, displayed and demanded by each individual unit of humanity will the people of this planet change the status quo and learn the true purpose for being given life.

It has been recorded by Baird T Spalding in his life's experience with "The Life And Teachings Of The Masters Of The Far East" published by De Vorss & co PO Box 550 Marina Del Rey, CA 90294 (these masters are the ones who have lived like you and me and who have risen to the true calling and become the Christ of themselves.) that the whole of the cosmic system consists of ninety one universes with a central sun ninety one thousand times larger than the ninety one universes. This

Central sun is known as the cosmic sun. From it comes the great white light which is the source of all created matter, it is GOD the Father It is from this point we will look at that which will follow.

On the vastness of the true although unknown heavenly bodies contemplate the words of Jesus when he said "In my Fathers kingdom their are many mansions, I go to prepare a place for you."

Think of that vastness, place yourself in it, how small and in significant are you, how insubordinate are you of the principles that control and govern this huge mass of time, space, and particles. Know you not that when you learn to control yourself and seek direction only from above you are guaranteed dominion over all these things.

Stop, think of the intelligence required to create all of the heavens plus the unshakeable law and principles that control it and make it and all it contains a continuous renewable system. How would your own ego compare yourself with this creation?

Now know this. You are just as much a part of this whole system as any other part, and what is more you have the same equal rights and above all responsibility to the great creative Principle as all other parts. Every part has a part to play and if that part is operating within the law of the Principle it is being of beneficial service to the whole. Even the planet earth and all its contents are in service for the wellbeing of the life forms thereon.

It is foolhardy and reckless to imagine that the continuous pollution and wasteful attitude of humanity against the perfect balance of the universal natural structure of the principle can be maintained without repercussions.

When things happen against the hand of man it is called an act of GOD, this is not a correct interpretation as the total overall perfect principle is not allowing any disruption to it's perfection and will correct any in perfection whichever way necessary.

Floods, storms, earthquakes, natural disasters, ice ages, even to the wiping out of nearly all humanity are the instruments available to principle to rebalance the perfection which holds the whole of the ninety one universes together. Principle is only protecting your inheritance; it is up to you when you collect it.

There has always been some confusion with regards the people of the light, the light people, (not the light workers) the crystal race. These are the people who were created by the pure white light vibration. How did this come about? Love is the prime cause of GOD the Father. God the Father is the creating light vibration. Firstly there had to be a building block on which to create the human form.

It has now been proven by New Zealand scientist Brian St Clair Corcoran at the International University in Los Altos, USA that lights electro magnetic field can be bent or vortexes to produce matter.

The bending of that light is that which is responsible for the creation of the Adamantine particle and as afore mentioned the Adamantine particle is the love upon which the human form has its total being, the beginning of life as we know it.

In light of the forgoing you will have found that the love of Jesus was for the spirit of man not just the flesh of humanity, other than the flesh is the vehicle for the spirit. Therefore his demonstrations of love was to serve the Divine Ideal Principle of actuality. That is to say he worshiped the law and principle which govern the principle of harmonious continuous sustainable renewable everything within the whole of the ninety one universes.

For the smallest particle of dust to the largest star or planet and every thing within including humanity have a dependency on one another for harmonic balance to keep the heavens as a service to all that is.

Jesus was of service to the spirit of all, planet earth is a service to all that live in on or around it even as it is being raped for the man made idea of profit and power over.

At this time Principle is actively engaged in rebalancing the harmony of planet earth to meet its commitment to Principle of harmony with the rest of the universes. You will show by your word and action how long you think you can deplete natures gift of environmental sustainability, the PRIME necessity of life, and get away with it?

How do you get on the path to your true inheritance? Firstly it is your own conscious expansion into that which is and has always been, not a mental surrender to another persons reason of thought It is a personal journey, for as you may walk with many others they in fact have no control over your mental activity, unless you surrender and adopt that other persons thought.

That being so it will be necessary to let go of all the thought other people have instilled in your mind. Use reason as only a process of thought not for establishing fact. Eliminate ego from all thought and activity. Use, think, and implement the attributes of the second wolf.

Renew you activity, thought and presentation of yourself inwardly and outwardly to bring it in accordance with the Divine Ideal Principle of The Universe. Have frequent conversations with the Father; be not afraid to ask him for help, if you are honest in your request to travel the path you will be surprised how willing he is to show the way.

The past is what has got you to your present situation, and if you stay with that, that is where you WILL stay. You will need to leave the club, association, tribe, national, religious, critical, judgmental thinking behind and see only the spiritual good in all things. By eliminating all the above it leaves the mental processes open to receive incoming information. See and activate only the positive, give the negative no thought, for your thought will give it power. Never be in doubt about principle, doubt will only slow your progress. Be aware other people are on the path and many are very much misguided by ego, for it is man's own egotistic arrogance, which compels him to be ignorant.

Forgive and forget the past and get on with your own conscious development.

Do not project your thinking into the future for it will develop as you expand your conscious thought and activity into the right use of the principle.

Bear in mind that Jesus came for the sole purpose of demonstrating the correct manner in which to make right use of the all embracing, continuous, sustainable, renewable, creative energy of the whole of the ninety one universes, Right –use—ness and to tell you that if you did it the way it was demonstrated the kingdom of heaven was yours. You would have arrived in your Father's house.

So certain was Jesus of the creative energy and the entire activity principle which governed it, that he lived and worshiped that and held it in the greatest reverence praying and giving his all to only the creative energy "GOD THE FATHER "

A point to ponder. IN the time of Jesus the language spoken by him was not Greek nor Latin, it was Aramaic, and that not having been taken into account in the translations has lead to the many incorrect interpretations of the Bible

Another point to ponder is that Jesus never wrote anything. It was only the interpretation of the listener which was recorded and the reinterpretations which have followed over the years that have given us the Bible of today.

One of the greatest fears of mankind has been brought about from the misunderstanding of reincarnation. If its true meaning was understood all would know that from the time of the creation of the souls, all have been through something near six or seven hundred incarnations and at least one life time in every civilization, every nationality, every religion, every creed, every belief system, every chance to be of service to your fellow man, every situation, to give the soul every opportunity to express itself within the law of the Divine principle.

You have been there and done that. Through the egotistic attitudes of the me we are by and large slow learners.

As the life force has been created by God with the preordained ultimate destiny to remember who we really are and claim dominion over all of God's creation it is important to know the pathway to that destiny.

For long enough we have been locked into the illusion of success, war. murder, violence, hate, fear, lies, corruption, religious division, and maintaining the man made systems of all the illusions, so it is time for the brave to begin the journey for home and remember our true destiny.

All this time we have been supporting the illusion created through man made law for financial wealth and control over. to the extent that our world and all it's systems are bringing the planet Earth to the point that for a reasonably long time it will not be able to support life as we know it.

The make up of Divine principle for the whole of the universe will not allow the incorrect application of human thought to divert any further the destruction of your space ship Earth.

Divine Principle is already at work warning of the result of the implementation of the illusion inflicted by incorrect action derived from incorrect thought. This is being shown us by the changing course of nature, endangered species, weather patterns, ice melt, rising temperatures, expanding deserts, lack of quality drinking water, plus all the other counter natural phenomena to let you know it is way past time to implement your highest thought and take care of that which God has created, for you to have life. You have not looked after it, why should you have it.

Man's egotistic thinking does not allow a breakthrough until he has a breakdown. For little does he realize that on the other side of crisis there is opportunity a plenty. When all else fails and no progress is being made, when more peoples die from hunger, disease and because of the implementation of system dictates compelling human beings to live in animal like conditions, then realization that he has come to a brick wall is that which is necessary for him to take stock of all situations and with new understanding start on a completely new direction.

Use all the experience gained from the breakdown or crisis, let go of the past, implement only positive thought with love for the benefit of all and enter boldly into your chosen activity.

The beauty of the return home is that it is for the individual only and all who walk with you can only advance according to their understanding gained through their own experience.

The truth of your being has been so terribly distorted that the illusion we live in has become our natural way of life, so here is the true destiny of all souls.

When you return to the Father's house these are some of the things you will be able to do. Like our great demonstrator JESUS has promised, "all these things you will do and more" you will be able to turn water into wine, have bread roll off your hand as long as there is a need, walk on water, heal the sick, return life to the dead, feed a multitude by producing from the aqueous mass, travel at the speed of thought, that means think of Mars and leave your body on earth, you are then on Mars with a body suitable for that environment, come back when ready and reenter the body you left behind, design a house in the mind and have it instantly appear, have no further use for it and return it to the aqueous mass from which it came or leave it to some one in dire need, if the mountain is in the way have it move instantly, walk through fire. be warm in frozen conditions, be clothed in what ever you can bring to mind, walk in the brilliance of the great white light. If you can think and act at this level of frequency it will happen as the law of supply is the promised abundance. "Before you have ask I have given."

Money yes. When you have dominion over the earth and can do the things mentioned above you will be able to hold the thought of the money you want in mind and it will roll off your hand until you dismiss the thought or you drown in money. It is purely a matter of mental manipulation of atomic structure through a fully developed mind.

If the ego of those who say they represent GOD can not do these things, who do they represent? Why do they need ten per cent of your earnings? If they do not know these things are possible what have they to teach? Are these the ones Jesus refereed to as false Gods or blind leaders of the blind?

You ask with all that power available and such strife on earth why does not God do something about it? He has. Before you have ask I have given.

GOD the Father the greater creator, has given you all and individually to ability to do it for yourself and have the world the way you want it by creating you in the image of GOD there by creating you in union with himself, and has given you the ability and choice of time to do it. You already have the power, step outside the illusion and make that time now.

Firstly you are a child of GOD a part of that family and therefore you are able to share all the things created by the Father for his family. You are in union with your Father.

Secondly your physical body has been created as a human begin with out any attachments of other peoples thoughts.

Third you are born into a piece of land created by God with all the means of sustaining life. This land has been given a name which is called by man a nation.

Forth you are totally exposed to the belief system of those around you and given the impression that those who advise are the ones who know the answer to your destiny.

Lastly you were created for God's purpose with a choice to follow and implement your OWN highest ideal. You have not been created to be a slave to the ideas of man.

A few tips to make it easy and accelerate your start. Think only POSITIVE. Speak only TRUTH. EXPECT nothing of others, they can not comply with your unknown expectations and only disappointment can result, they are on their own path. ASS U ME (assume) nothing as it will make an ASS of U and ME. Show others how to do, not do it for them as you will rob them of the EXPERIENC. JUDGE not, you do not know their soul agenda. Think and talk not MALICE of others, its effect will return to you. If ill is spoken OF you return only LOVE. In place of JUDGMENTS of self make DECISIONS. Do not CRITICIZE, make and speak only of your OBSERVATIONS based on what you know from experience, not your belief system. In all your decision making regarding OTHERS, use the wisdom of Solomon, a fair and evenly balanced win win situation. Use your position in the community wisely, 'Consider the wealth of Solomon', reincarnation always repays in like kind x 4. As a representative use total integrity total honesty total truth and total openness.

To get yourself motivated sit quietly, completely let go of your ego, contemplate GOD, say the name GOD very meaningfully, consider your highest thought in line with Divine Principle, now with all the honesty, integrity, and truth you can bring into being, implement that into your daily routine until it becomes part of your natural way of thinking, saying, and doing.

While you are implementing your first high thought you can start work on the next high thought and the next and so on. At each step your vibratory rate will increase, life will take on a new meaning, things will start to rearrange themselves and even higher thoughts will present themselves for your enlightenment consideration and presentation.

Implement all your high thoughts with all the vigor and love you can call to your aid.

Most important give thanks for all your blessings, your progress, and pray for love to envelope the world blessing all within.

Writing of this nature is a guide only, it is not an experience, knowing it from all the books is not the true knowledge. The ideal thought put into action by yourself brings the experience, that is the knowledge and the accumulation of the experiences you have creates your own wisdom.

Without judgment, that which you consider are other peoples mistakes are the negative experiences you do not need to have.

Ego is the thought which imagines that the me and its thought is greater than the DIVINE PRINCIPLE OF THE UNIVERSE, THE, I AM

Consequently observation shows that It is mans own egotistic arrogance which compels him to be ignorant of the reason he has life and the destiny that has been preordained by the creator.

The last two thousand years of misinformation has lead everyone to believe that at some time in the future some divine person or action will come and lift those who think they are chosen or have lived a good life into that which we call heaven.

There are those who write, talk, and believe of light workers, grid fixers, galactic federation along with other beliefs which is all part of the system of encouraging everyone to wait for some other outside agency to come and do the work for them. IT CAN NOT HAPPEN! He who hesitates is lost.

Jesus said "None will come to the Father but by me" meaning you will not have the ability I have unless you follow and implement my instructions of love and good will to all. 'The attributes of the second woof'

The only uplifting into heaven is when you yourself and I will say that again YOU YOURSELF lift your own vibration through the implementation from the inner core of your being, give love to the whole of the universe and all within.

The second coming of the Christ is when you become the CHRIST of yourself.

You and only you can lift your thought word and action to the higher plane thereby lifting your vibratory rate to that of the vibratory rate of the Father, The house of the Father, home and the Christ of yourself.